AF539716

Diseases of Field & Horticultural Crops & Their Management

NIPA® GENX ELECTRONIC RESOURCES & SOLUTIONS P. LTD.
New Delhi-110 034

About the Author

Dr. Sanjeev Kumar is Assistant professor/Scientist, Department of Plant Pathology, Jawaharlal Nehru Krishi Vishwavidyalaya, Jabalpur, Madhya Pradesh. He is very sincere and hard working teacher. He has taught many undergraduate, postgraduate and Ph.D courses with great dedication to the great satisfaction of the students and actively involved in all teaching activities. Dr. Kumar has guided twenty five M.Sc and two Ph.D student and has been actively involved in his research and extension activities too. As a researcher, he has been associated with 8 major research projects sponsored by ICAR, NATP,TSP and JICA, Japan. He has published 68 research and reviews papers, 25 book chapters, 105 popular articles, 1 Technical Bulletin and 3 Technical Folders. He is an active member of over six National and International Societies. He is the fellow member of Indian Phytopathological Society, IARI New Delhi, Indian Society of Mycology and Plant Pathology, Udaipur, Rajasthan and Society of Biocontrol Advancement, Banguluru. He has attended many conferences, symposia, workshops and presented papers. He has also attended the World Soybean Research Conference held in Durban, South Africa.

Dr. Kumar has authored ten books entitled Plant Pathogens & Principles of Plant Pathology, Diseases of Horticultural Crops: Identification & Management, Diseases of Field Crops and Their Integrated Management, Pesticides and Plant Protection Appliances, Fundamentals of Plant Pathology, Integrated Disease Management: Principles & Practices, Principles of Plant Pathology, Diseases of Field &Horticultural Crops & Their Management-II published by New India Publishing Agency, Pitam Pura, New Delhi, Diseases of Field &Horticultural Crops & Their Management-I, published by Brillion Publishing Desh Bandhu Gupta Road, Karol Bagh, New Delhi – 110005 and Principles of Plant Disease Management, published by Kalyani Publishers, Daryaganj, New Delhi and three practical manuals. His research interests include pulse pathology and biological control.

2nd Fully Revised and Enlarged Edition

Diseases of Field & Horticultural Crops & Their Management

Sanjeev Kumar
Assistant Professor/Scientist
Department of Plant Pathology
Office of Dean, Faculty of Agriculture
Jawaharlal Nehru Krishi Vishwa Vidyalaya
Jabalpur-482 004, Madhya Pradesh, India

NIPA® GENX ELECTRONIC RESOURCES & SOLUTIONS P. LTD.
New Delhi-110 034

NIPA® GENX ELECTRONIC RESOURCES & SOLUTIONS P. LTD.

101,103, Vikas Surya Plaza, CU Block
L.S.C. Market, Pitam Pura, New Delhi-110 034
Ph : +91-11-43860225, Mob.: +91 9717133558, 9540816132
E-mail: newindiapublishingagency@gmail.com
Website: www.nipaersources.com

Print ISBN: 978-93-58874-34-1

ebook ISBN: 978-93-58874-31-0

Disclaimer: The views expressed in the articles including the contents are sole responsibility of the respective authors. The editors bear no responsibility with regards to source, claims made and authenticity of the contents.

NIPA® also publishes books in a variety of electronic formats. Some content that appears in print may not be available in electronic books, and vice versa.

Composed and Designed by NIPA®.

Dedicated to
My Parent

Preface

The Sixth Deans' Committee Recommendations have led to a thorough revision of the B.Sc (Hons) Ag, Plant Pathology program and has rendered existing text books incomplete for students. The course content has to be rewritten to comply with the NEP-2020 requirements for the new academic framework. This book of Diseases of Field Crops & Horticultural Crops & their Management is carefully restructured for undergraduate class and is completely in accordance with the latest 6th Deans's Committee Recommendations Guidelines.The objective of writing this book are to study the symptoms produced on the host, study the etiology of the diseases, know about the disease cycle of the pathogens during pathogenesis, study the epidemiological factors responsible for disease development and study the management techniques for curbing the major diseases of field and horticultural crops. The main features of this volume are as follows:

- The chapter "Fun Time" will sets the tone by addressing general terminology which also relate to the subject. Fun time is good ice breaker and prepares the students for the information to follow .
- The subject matter has been presented point wise and step by step with suitable examples.
- Efforts have been made to provide recent information.
- The book has 39 chapters that cover the entire syllabus of course at the undergraduate level in Agriculture universities.
- Chapters are separated into two categories: diseases of field crops, such as rice, wheat, maize, sorghum, bajra, finger millet, groundnut, soybean, pea, black gram and green gram, sugarcane, mustard, sunflower, cotton and diseases of horticultural crops, such as citrus, papaya, pomegranate, apple, peach, grapevine, strawberry, coconut, tea, coffee, mango, potato, tomato, brinjal, chilli, cucumber, cruciferous vegetables, beans, okra, turmeric, coriander, rose, and marigold.
- Each chapter has further be divided into sub-heads such as important diseases and their causal organisms, diagnostic symptoms on different parts, etiology, disease cycle, epidemiological conditions and integrated management.

- Management practices including, cultural, physical, biological, host resistance and need based chemicals leading to integrated disease management.
- The topic and the contents has been arranged in such a manner that students get an easy reading with due continuity and the understanding.
- Each chapter ends with a sample question paper. Sample papers contain all of the syllabus's important topics and frequently answered questions, making them thorough revision aids. Frequent use of them strengthens knowledge, sharpens reasoning abilities, and dispels uncertainty, resulting in a more sophisticated comprehension of the material.
- The subject index has been provided at the end for easy search of the information.
- This book wold also be useful to graduate and post graduate students appearing for competitive examination like SRF, JRF, NET & ARS.
- This book will serve immensely for teachers and students in the field of plant pathology.

I do not claim originality in the preparation of this book and has taken help from a large number of books, journals, periodicals, bulletins, google etc. I sincerely thank the publishers, editors, and writers of the books, magazines, etc. In addition, I owe my children, Saumya and Adyan, for putting up with me as I prepared this book, and my wife, Dr. (Mrs.) Archana Rani, for her support and inspiration.

I hope the book inspires students to go on a quest where asking questions leads to greater learning than just being taught. We would be happy to include recommendations and enhancements made by educators and learners in further editions.

Jabalpur **Sanjeev Kumar**

Contents

Key Terms (Fun Time)

Term	Definition
Abiotic	Nonliving, or caused by a nonliving agent; e.g., abiotic disease.
Acervulus	A subepidermal, saucer-shaped, asexual fruiting body producing conidia on short conidiophores.
Active ingredient	In pesticides, the chemical responsible for the desired effect.
Aecia	Rust fruiting bodies that produce aeciospores.
Aeciospores	Dikaryotic spore of a rust fungus produced in an aecium; in heteroecious rusts, a spore stage that infects the alternate host.
Aggressiveness	Virulent forms of pathogen cause differing degrees of symptom severity.
Alternate host	One of two kinds of plants on which a parasitic fungus (e.g., rust) must develop to complete its life cycle.
Anamorph	Imperfect or asexual stage of a fungus.
Antheridium	Male sexual organ found in some fungi.
Anthracnose	A disease that appears as black, sunken, leaf, stem, or fruit lesions, caused by fungi that produce their asexual spores in an acervulus.
Antibiotic	A chemical compound produced by one microorganism that inhibits or kills other microorganisms.
Apothecia	An open cup- or saucer-shaped ascocarp of some ascomycetes.
Apressorium	Swollen tip of a hypha or germ tube that facilitates attachment and penetration of the host by a fungus.
Aquisition period	Period of time need for a virus vector to aquire a virus while feeding on an infected plant host.
Ascocarp	The fruiting body of ascomycetes bearing or containing asci.
Ascomycetes	Fungi that produce sexual spores in sac-like structures called asci.
Ascus	A sac-like cell of a hypha in which meiosis occurs and that contains ascospores (usually eight).
Asexual reproduction	Any type of reproduction not involving the union of gametes or meiosis.
Autoecious	Rusts that complete their life cycle on a single host.
Autoecious fungus	A parasitic fungus that can complete its entire life cycle on the same host.
Basidiospores	Sexual spores of basidiomycetes prodcued on basidia.
Biological control	Total or partial inhibition or destruction of pathogen populations by other organisms.
Biotic	Living; associated with or caused by a living organism.
Biotroph	An organism that can live and multiply only on another living organism.
Blight	A disease characterized by general and rapid killing of leaves, flowers, and stems.
Canker	A necrotic, often sunken, lesion on a stem, branch, or twig of a plant.

Chlamydospores	A thick-walled asexual spore formed by the modification of a hyphal cell.
Chlorosis	Yellowing of normally green tissue due to chlorophyll destruction or failure of chlorophyll formation.
Cleistothecia	An entirely closed ascocarp.
Coat Protein	Protein, encoded by virus nucleic acid, which covers the viral nucleic acid
Coenocytic hyphae	Hyphae without crosswalls, found in Oomycetes
Conidia	Asexual, non-motile spores of fungi.
Conidiophore	A specialized hypha that produces conidia.
Curative	Chemical control method aimed at inhibiting the development of an established infection.
Curl	Abnormal leaf or shoot bending or curling brought on by localized overgrowth in a particular tissue or on one side.
Damping-off	Destruction of seedlings near the soil line, resulting in the seedlings falling over on the ground.
Defoliation	Loss of leaves through abscission.
Dieback	Progressive drying, shrivelling, and browning of twigs or branches from the tips inward toward the trunk.
Disease	Any malfunctioning of host cells and tissues that results from continuous irritation by a pathogenic agent or environmental factor and leads to development of symptoms.
Disease cycle	The chain of events involved in disease development, including the stages of development of the pathogen and the effect of the disease on the host.
Disease incidence	Number or poroportion of individuals that are diseased.
Disease severity	Amount or proportion of tissue that is diseased.
Downy mildew .	A plant disease in which the sporangiophores and spores of a fungus appear as a downy growth on the lower surface of leaves and stems, fruit, etc., caused by fungi in the family Peronosporaceae.
Dwarfing	Subnormal size of a plant or some of its organs.
Ectoparasite	A parasite feeding on a host from the exterior.
ED 50	Effective dose, the amount needed to have the desired effect in 50% of the population.
Endoparasite	A parasite that enters a host and feeds from within.
Epidemic	A disease increase in a population; usually a widespread and severe outbreak of a disease.
Epidemiology	Study of factors affecting the outbreak and spread of disease.
Epiphytotic	A widespread and destructive outbreak of a disease of plants; epidemic.
Fission	Bacterial reproduction by simple cell division.
Fumigant	A toxic gas or volitile substance that is used to disinfest soil of various pests.
Fungicide	A compound toxic to fungi.
Fungistatic	A compound that prevents fungus growth without killing the fungus.
Gall	A noticeable tumefaction that is typically made up of undifferentiated cells and is frequently spherical.
Germ tube	The early growth of mycelium from a germinating spore.

Girdling	A canker's tangential expansion or the lateral coalescence of cankers, which encircles a branch or stem and completely stops its conduction.
Gram negative	Group of bacteria with two membranes in association with their cell wall.
Gram positive	Group of bacteria with one membrane in association with their cell wall.
Gummosis	Formation of gums by diseased cells and tissues and the extrusion of gum from wounds and other lesions.
Haustorium	A specialized fungal hyphae that enters and absorbs nutrients from a host cell.
Hemibiotrophic	An organism that lives part of its life as a parasite on another organism and the other part as a sarophyte.
Heteroecious	Rusts that require two different hosts to complete their life cycle.
Hyperparasite	A parasite parasitic on another parasite.
Hyperplasia	A plant overgrowth due to increased cell division.
Hypertrophy	A plant overgrowth due to abnormal cell enlargement.
Hypha	Vegetative structure of fungi and Oomycetes.
Inclusion body	Aggregation of virus partilces in a host cell that is visible with a compound microscope.
Incubation period	The period of time between penetration of a host by a pathogen and the first appearance of symptoms on the host.
Infection	The establishment of a pathogen within a host plant.
Injury	Damage of a plant by an animal, physical, or chemical agent.
Inoculation period	Period of time need for a virus vector to transmit a virus while feeding on an uninfected plant host.
Inoculum density	Number of infective units in a given volume or area.
Integrated control	An approach that attempts to use all available methods of control of a disease or of all the diseases and pests of a crop plant for best control results but with the least cost and the least damage to the environment.
Latent period	Period of time before a vector is infective after picking up the virus from an infected host.
Lesion	A localized area of diseased tissue.
Life cycle	The stage or successive stages in the growth and development of an organism that occur between the appearance and reappearance of the same stage (e.g., spore) of the organism.
Local lesion	A localized spot produced on a leaf upon mechanical inoculation with a virus.
Macrocyclic	Rusts that produce all five spore types.
Mechanical transmission	Virus transmission from plant to plant by infected plant sap.
Microcyclic	Rusts that lack one or more of the five spore types.
Mode of action	Molecluar mechanism of a pesticide; how the chemical interacts with the pathogen. Pesticides are grouped by mode of action. Some groups of pesticides are more likely to lead to the development of resistance than others.
Monocyclic disease	A disease where only one disease cycle is completed each year or growing season.

Monocyclic	Having one cycle per season.
Mosaic	Symptom of certain viral diseases of plants characterized by intermingled patches of normal and light green or yellowish color.
Mycelium	The hypha or mass of hyphae that make up the body of a fungus.
Necrotic	Dead and discolored.
Nectrotroph	A microorganism feeding only on dead organic tissues.
Obligate parasite	A parasite that in nature can grow and multiply only on or in living organisms.
Oogonium	Female gametangium of oomycetes containing one or more gametes.
Oospore	A sexual spore produced by the union of two morphologically different gametangia (oogonium and antheridium).
Ooze	Viscid masses composed of living or dead pathogen structures and partially disintegrated host tissues.
Pathogen	An entity that can incite disease.
Pathogenicity/ Virulence	Ability of a microbe to cause disease (invade, infect, cause symptoms, reproduce).
Perithecia	A flask-shaped ascocarp with an opening for releasing spores.
Plasmid	Small, circullar, extrachromosomal DNA found in bacteria.
Plasmodesmata	A connection across a plant cell wall that connects the cytoplasm of two neighboring cells.
Polycyclic	Completes many (life or disease) cycles in one year.
Polycyclic disease	A disease with several cycles each year of inoculum production and infection.
Polyetic	Requires many years to complete one life or disease cycle.
Preventative	Chemical control method aimed at preventing infection of the pathogen.
Primary infection	First infection of a plant by the overwintering or oversummering pathogen.
Primary inoculum	Overwintering or oversummering pathogen, or its spores that cause primary infection.
Primary inoculums	Inoculum that produces the first infection of plants in a year or growing season.
Pustule	Small blister like elevation of epidermis created as spores form underneath and push outward.
Pycnidium	A flask-shaped asexual fruiting body with an opening for releasing spores..
Race	A genetically and often geographically distinct mating group within a species; also a group of pathogens that infect a given set of plant varieties.
Resistant	Phenotypic expression related to complete or partial suppression of symptom severity and/or pathogen reproduction and accomplished by arrested or slowed invasion of host by pathogen.
Resting spore	Long term survival spores produced by Plasmodiophoromycetes.
Rust	A disease giving a "rusty" appearance to a plant and caused by one of the Uredinales (rust fungi).
Sanitation	The removal and burning of infected plant parts, decontamination of tools, equipment, hands, etc.
Saprophyte	An organism that uses dead organic material for food.

Scab	A limited, more or less circular, raised, and sometimes roughened lesion on fruits, tubers, leaves, and stems resulting from an overgrowth of epidermal, cortical, and peridermal tissues.
Sclerotia	Macroscopic mass of hyphae, usually rounded and darkened.
Secondary infection	Any infection caused by inoculums produced as a result of a primary or a subsequent infection; an infection caused by secondary inoculum.
Secondary inoculum	Inoculum produced by infections that take place during the same growing season.
Septate	Having cross walls in a hypha or spore. A cross wall is called a septum.
Sign	Portions or byproducts of the pathogen that are visible on the host plant. A sign is the pathogen itself.
Soil transients	Parasitic microorganisms that can live in the soil for short periods.
Sporangiophore	A specialized hyphae that produces sporangia.
Sporangium	Non-motile asexual spore produced by oomycetes, lemon or globulose in shape.
Spore	The reproductive unit of fungi consisting of one or more cells; in function, it is analogous to the seed of green plants.
Sporodochium	An asexual fruiting structure consisting of a cluster of conidiophores woven together on a mass of hyphae.
Susceptible	Phenotypic expression related to extensive symptom development and/or pathogen reproduction and accomplished by uninhibited invasion of host by pathogen.
Symptom	Abnormal plant function or growth brought on by a disease. A symptom is the plant's reaction.
Synnema	An asexual fruiting body consisting of fused conidiopores to form a stalk with conidia on the end.
Systemic	Chemical meant to be taken up by and distributed throughout the plant as a preventative or curative control measure.
Teleomorph	Perfect or sexual stage of a fungus.
Teliospores	Overwintering spores of rusts and smuts that produce basidia and basidiospores.
Tumor	Local swelling on any part of the plant, usually woody roots, stem, or branches, usually resulting from stimulation of the plant meristem by the pathogen.
Uredia	Rust fruiting bodies that produce uredospores.
Uredospores	Asexual, dikaryotic, often rusty-colored spore of a rust fungus, produced in a structure called a uredinium; the "repeating stage" of a heteroecious rust fungus, i.e. capable of infecting the host plant on which it is produced.
Vector	Disease transmitting agents.
Vector transmission	Transmission of a pathogen from plant to plant by another organism.
Viroids	Small RNAs that can infect a host and cause disease.
Virus	A submicrscopic obligate parasite consisting of nucleic acid and protein.
Wart	A horny, hardened protuberance.

Wilt	Loss of rigidity and drooping of plant parts, generally caused by insufficient water in the plant.
Yellowing	Loss of green color from chlorophyllous tissues, due to the destruction of the chlorophyll .
Zoospore	A motile asexual spore characterized by two flagella.

1

Rice Crop Diseases & Management

Major Diseases	**Causal Organism**
Blast	*Pyricularia oryzae*
Brown spot	*Helminthosporium oryzae*
Sheath blight	*Rhizoctonia solani*
False smut	*Ustilaginoidea virens*
Bacterial leaf blight	*Xanthomonas oryzae pv. oryzae*
Bacterial leaf treak	*Xanthomonas oryzae pv. oryzicola*
Tungro	*Rice tungro bacilliform virus* and *Rice tungro spherical virus*
Khaira	Zn Deficiency

1. Blast

Diagnostic Symptoms

- Pathogen attacks the crop at all stages of crop growth.
- Symptoms appear on leaves, nodes, rachis, and glumes. On the leaves, the lesions appear as small bluish green flecks, which enlarge under moist weather to form the characteristic spindle shaped spots with grey centre and dark brown margin (**Leaf blast).**
- Spots coalesce as the disease progresses and large areas of the leaves dry up and wither. Severely infected nursery and field appear as burnt.
- Black lesions appear on nodes girdling them. The affected nodes may break up and all the plant parts above the infected nodes may die **(nodal blast).**
- During flower emergence, the fungus attacks the peduncle and the lesion turns to brownish-black which is referred to as rotten neck / neck rot / panicle blast **(neck blast**).
- In early neck infection, grain filling does not occur while in late infection, partial grainfilling occurs (Fig.1).

Etiology

Scientific Classification

Kingdom : Fungi

Division : Eumycota

Sub- division : Deuteromycotina

Class : Hyphomycetes

Order : Hyphomycetales (Moniliales)

Family : Dematiaceae

Genus : *Pyricularia*

Species : *oryzae*

- Anamorphic stage: *Pyricularia oryzae*
- Telemorphic stage: *Magnaporthe grisea*)
- Mycelium is hyaline to olivaceous and septate.
- Conidia are produced in clusters on long septate, olivaceous conidiophores.
- Conidia are pyriform to ellipsoid, attached at the broader base by a hilum.
- Conidia are hyaline to pale olive green, usually 3 celled.
- Perfect state of the pathogen is *M. grisea* producing perithecia.
- Ascospores are hyaline, fusiform, 4 celled and slightly curved.

Disease Cycle

- Primary infections generally result in a few widely scattered spots on leaves. Mycelium and conidia in the infected straw and seeds are major sources of inoculums.
- Pathogen also survives on collateral hosts *viz., Brachiaria mutica, Leersia hexandra, Panicum repens, Digitaria marginata,* and *Echinochloa crusgalli.*
- Spores arising from the primary infections are capable of causing many more infections. This cycling is called secondary spread.
- Secondary spread is responsible for the severe epidemics of blast in fields and localized areas.

- Disease spreads primarily through airborne conidia since spores of the fungus present throughout the year.

Epidemiology

- High relative humidity -90-99 per cent.
- Low night temperature -between 15-20°C or less than 26°C.
- Cloudy weather, Intermittent drizzles, more of rainy days, longer duration of dew.
- Excess dose of nitrogen and availability of collateral hosts.

Management

- Grow resistant to moderately resistant varieties ADT36, ADT39, CO47, IR 20, ASD 18 and IR64.
- Remove and destroy the weed hosts in the field bunds and channels.
- Treat the seeds with Carbendazim 50 WP or Tricyclazole75 WP at 2 g/kg or *Pseudomonas fluorescens* @ 10g/kg of seed.
- Early sowing of seeds after the onset of the rainy season is more advisable..
- Excessive use of fertilizer should be avoided as it increases the incidence of blast. Nitrogen should be applied in small increments.
- Water management practices in rain fed areas lessen the likelihood of stress, which also aid in blast control.
- Spray the crop with Edifenphos or Carbendazim or Tricyclazole @ 0.1%, on appearance of the disease. Four to five sprays at 10–15 days interval may be needed for complete control.

2.Brown Spot

Diagnostic Symptoms

- Initial symptoms appear as minute circular or oval, brown lesions or dots on the coleoptiles, leaf blade, leaf sheath and glumes, being most prominent on the leaf blade and glumes.
- These lesions may girdle the coleoptiles and cause distortion of the primary and secondary leaves leading to seedling blight.
- Spots have a light brown to gray centre, surrounded by a reddish brown margin.

- Several spots coalesce and the leaf dries up.
- Infected seedlings become stunted or die.
- Affected nurseries can be often recognized from a distance by their brownish scorched appearance.
- Velvety appearance of lesions is visible on infected glumes under severe conditions.
- Infected grains show black discoloration or brown lesions.
- Disease-causing fungi can also penetrate grains, causing *'pecky rice',* a term used to describe spotting and discoloration of grains.
- Disease appears from seedling to milking stage(Fig. 2).

Etiology

Scientific Classification

Kingdom : Fungi

Division : Eumycota

Sub- division : Deuteromycotina

Class : Hyphomycetes

Order : Hyphomycetales (Moniliales)

Family : Dematiaceae

Genus : *Drechslera*

Species : *oryzae*

- Anamorphic stage: *Drechslera oryzae*
- Teleomorphic stage: *Cochliobolus miyabeanus*
- Pathogen grows both intra and intercellularly inside the host.
- Mycelium brown in color and produces conidiophores that emerge in group by rupturing epidermis or through stomatal openings.
- Conidiophores arise singly or in small groups, geniculate and brown in colour.
- Conidia are usually curved with a bulged center and tapered ends.They are pale to golden brown in colour and are 6-14 septate.

- Conidia germinate readily producing germ tube mostly from cells of both ends.
- Perithecia with asci containing 6-15 septate, filamentous or long cylindrical, hyaline to pale olive green ascospores.

Disease Cycle

- Pathogen perennates by infected seeds and collateral hosts growing to the adjacent to paddy fields.
- Fungus also survives on collateral hosts like *Leersia hexandra* and *Echinochloa colonum* and *Cynodon dactylon* etc.
- Infected seeds and stubbles are the most common source of primary infection.
- When such seeds are sown in nurseries, the pathogen results in seedling blight.
- Similarly ,condia produced on infected collateral hosts get disseminated by air and causes primary infection on seedlings.
- Secondary infection takes place with air and water borne conidia, infect the plants both in nursery and in main field.

Epidemiology

- Temperature -25-30°C.
- Relative humidity -above 80 per cent.
- Excessive application of nitrogen.

Management

- Removal of collateral hosts and infected debris from the field.
- Use of slow release nitrogenous fertilizers.
- Use disease free seeds.
- Seed treatment with hot water (53−54°C) for 10−12 minutes before planting, to control primary infection at the seedling stage. To increase effectivity of treatment, pre-soak seeds in cold water for 8 hours.
- Seed treatment with Thiram or Captan at 4 g/kg.
- Foliar spray of nursery with Edifenphos (0.1%) or Mancozeb (0.25%) .

- Foliar spray the crop in the main field with Edifenphos 500 ml or Mancozeb 2 kg/ha, If needed repeat after 15 days.

3. Sheath Blight

Diagnostic Symptoms

- Initial symptoms are noticed on leaf sheaths near water level.
- On the leaf sheath oval or elliptical or irregular greenish grey spots are formed.
- As the spots enlarge, the centre becomes greyish white with an irregular blackish brown or purple brown border.
- Lesions on the upper parts of plants extend rapidly coalesing with each other to cover entire tillers from the water line to the flag leaf.
- Presence of several large lesions on a leaf sheath usually causes death of the whole leaf, and in severe cases all the leaves of a plant may be blighted.
- Infection extends to the inner sheaths resulting in death of the entire plant.
- Plants heavily infected in the early heading and grain filling growth stages produce poorly filled grain, especially in the lower part of the panicle.
- Pathogen affects the crop from tillering to heading stage(Fig.-3).

Etiology

Scientific Classification

Kingdom : Fungi

Division : Eumycota

Sub- division : Deuteromycotina

Class : Hyphomycetes

Order : Agonomycetales

Family : Agonomycetaceae

Genus : *Rhizoctoniza*

Species : *solani*

- Anamorphic stage: *Rhizoctonia solani* Kuhn
- Teleomorphic stage: *Thanetophorus cucumeris*
- In India, sheath blight of rice is caused by anastomosis group-1 (AG-I-1) of the fungus having 3-16 nuclei per cell.
- Mycelial stage producing sclerotia is the commonly encountered stage in India.
- Pathogen produces a fast growing stout mycelium with septate hyphae.
- Young colonies may look white but older mycelium is invariably some shade of brown.
- Sclerotia may or may not formed. In some isolates the sclerotia are light brown while in some they are so dark as to look black.
- In some isolates the mycelium produces barrel shaped cells which have been termed as chlaymydospores.
- Cells are as a rule multinucleate (in other species of Rhizoctonia they are binucleate).
- Sclerotia of the fungus are distinct from many other fungi in that there is no differentiation of the sclerotial tissue into a rind and internal medulla although outer cells may be darker and thicker walled.
- Basidial stage is mostly saprophytic stage of the pathogen. It appears as a flaky pellicle on the surface of the leaves near the ground level or dead parts of the stem.
- Basidia are barrel shaped , obpyriform or obclavate or clavate and measure 9-25x5-12 μm, bear 4 sterigmata.
- Basidiospores are ellipsoid or oblong-ellipsoid, flattened on one side, hyaline, truncate, and measure 5-15x 4-8 μm.

Disease Cycle

- Pathogen is seed borne. In addition to the weed hosts and infected seed, the pathogen mainly survives as sclerotia and /or mycelia in diseased plant debris left in the field.
- Sclerotia and mycelia which survive in plant debris are brought up on the soil surface during puddling, leveling, and other operations.
- They come in contact with new planted seedlings, germinate and cause infection.

Epidemiology

- High temperature -30-32°C.
- High relative humidity -96-97 per cent.
- Closer planting.
- Heavy doses of nitrogenous fertilizers.

Management

- Deep ploughing in summer and burning of stubbles.
- Grow resistant varieties.
- High seeding rate or dense plant spacing should be discouraged .
- Need-based or balanced application of nitrogen fertilizers in the fields.
- Sanitation, specifically removing of weeds, infected stubbles or crop residues from the field .
- Irrigation should be restricted in the affected crop and avoid flow of irrigation water from infected fields to healthy fields.
- Foliar spray with Hexaconazole 5% SC @1 ml/l or Carbendazim @1g/l or Propiconazole @ 1 ml/ 1 of water, if required repeat after 15 days.
- Many combination products like Trifloxystrobin and Tebuconazole combination @ 0.4 g/l, or Tricyclazole and Propiconazole combination or Flusilazole and Carbendazim combination have also been found very effective against the disease.

False Smut

Diagnostic Symptoms

- False smut is one of the emerging fungal grain diseases of rice.
- Symptoms are visible after flowering only, where the fungus infects the young ovary of individual kernel and transform them into large, velvety green balls (smut balls).
- Initially the smut balls are small and remain confined between glumes. They gradually enlarge and enclose the floral parts.
- Smut balls are initially yellow in colour and are covered by membrane which later bursts and the colour changes to orange, yellowish green, green, olive green and finally greenish black(Fig.-4).

- Only few grains in a panicle are usually infected and the rest are normal.

Etiology

Scientific Classification

Kingdom : Fungi

Division : Eumycota

Sub- division : Ascomycotina

Class : Plectomycetes

Order : Clavicipitales

Family : Clavicipitaceae

Genus : *Claviceps*

Species : *oryzae – sativae*

- Anamorphic stage*: Ustilaginoidea virens*
- Teleomorphic stage: *Claviceps oryzae - sativa*
- Mycella filling in the ovary are united, fine and colorless and they develop conidial. Pseudosclerotial and sclerotial stages.
- Conidia are spherical, echinulate, olivaceous in color, and measure 4-6x25µm.
- Conidia germinate in water giving rise to short germ tube, 1-3 secondary conidia are produced on each germ tube usually after 12-24 hours.
- Secondary conidia are subglobose or globose to oblong, hyaline, granulate, and measure 4-8x2-5µm.
- Late in the season, the conidial masses in smut balls harden and form pseudosclerotia.
- Sclerotia are formed in each green ball.The sclerotia are hard, variously shaped, and 5-13x2-5 mm in size.It germinate to produce stipes bearing perithecial stroma.
- Perithecia are ovate to pyriform measuring 150.5 to 430x86-193.5µm.
- They bear hyaline, elongated,cylindrical and each containing eight ascospores.
- Ascospores are hyaline, unicellular; germinate in water giving rise to 1-2 germ tubes, which bear secondary conidia.

Disease Cycle

- Sclerotia are reported to be the perennating structures.
- Sclerotia are considered to be the major source of primary inoculum.
- Ascospores produced inside the perithecia embedded in sclerotial stroma are disseminated and brought on to the host surface whereupon they germinate and infect the host just before the appearance of ears.
- Primary inoculums also come from the collateral hosts.
- It is considered however, that conidia play an important role in causing secondary infection during the growing season.

Epidemiology

- High relative humidity ->90%.
- Temperature ranging - 25–35 °C.
- Rain
- Soils with high nitrogen content.

Management

- Use healthy disease free seeds for sowing.
- Seeds should not be taken from false smut affected fields.
- Early planting can help to escape heavy disease incidence.
- Infected plants should be removed and destroyed at the time of harvesting.
- Field bunds and irrigation channels should be kept clean.
- Foliar spray of Carbendazim or Copper oxychloride fungicide @ 2.5–3.0g/litre of water at tillering and pre-flowering booting stages.
- Foliar spray of Propiconazole 500ml/ha or Hexaconazole 1000 ml/ ha before seed formation.

Bacterial Leaf Blight

Diagnostic Symptoms

The symptoms of this disease occur in two phases viz., leaf blight phase and seedling blight or 'kresek' phase.

Leaf Blight Phase

- Initially water-soaked lesions formed on the tip of the leaves which increase in length and width downwards and turn yellowish to straw colour stripes with a wavy margin.
- These lesions may develop at one or both edges of the leaves or along the mid rib.
- There is an appearance of bacterial ooze that looks like milky or opaque dewdrop on young lesions early in the morning.
- Lesions turn yellow to white as the disease progresses.
- Severely infected leaves tend to dry quickly.

Kresek Phase or Seedling Wilt

- Green water-soaked layer along the cut portion or leaf tip of leaves are early symptoms.
- Leaves wilt and roll up and become greyish green to yellow.
- Entire plant wilts completely.
- Infected leaves of mature plant turn yellow to pale yellow.
- Youngest leaf is uniform pale yellow or has broad yellow stripe.
- Under severe conditions., panicles remain sterile and unfilled but not stunted.
- Kresek phase is observed 1-3 weeks after transplanting.
- The kresek infected tillers may be confused with the stem borer injury but the latter can be easily pulled out while it is not so with the kresekinfected tillers(Fig.-5).

Quick diagnosis

- Cut a young lesion across and place in a transparent glass container with clear water.
- After a few minutes, hold the container against light and observe for thick or turbid liquid coming from the cut end of the leaf.

Etiology

Scientific Classification

Kingdom : Prokaryotae

Division : Gracilicutes

Class : Proteobacteria

Family : Psudomonadaceae

Genus : *Xanthomonas*

Species : *campestris pv. oryzae*

- Causal organism *: Xanthomonas oryzae pv. oryzae.*
- Rod shaped cells of the bacterium measure 0.5-0.8 x 1.0-2.0μ.
- Single and not formed capsule.
- Gram negative.
- Motile by a single polar flagellum.
- Colonies are slow growing, mucoid and straw colored to yellow.

Disease Cycle

- Pathogen survives in soil , seed, infected stubbles and collateral hosts.
- Seed from infected crop, diseased wild rice growing in ponds and Irrigation water contaminated with the bacterium flowing through field to field a provides the primary inoculum.
- Secondary spread is through wounds and stomata by bacterial cells disseminated by wind borne raindrop splashes, by irrigation water or rainwater coming from infested fields and by contact between diseased and healthy leaves.

Epidemiology

- Temperature -25-30^0 C.
- Heavy rain, heavy dew, flooding, deep irrigation water.
- Severe wind.
- Application of excessive nitrogen.

Management

- Use of healthy disease free certified seed .
- Grow resistant cultivars like IR 20 and TKM 6.
- Seed treatment with Streptocycline (1g) + Carbendazim 50 W.P. (20g) for 8–10 kg of seed in 10 litres of water for 12 – 15 hours .
- Hot water seed treatment (53–54°C) for 30 minute..
- Application of stable bleaching powder @12.5 kg/ha .
- Moderate level of 60 – 80 kg N /ha with required potassium may be recommended in endemic area during wet season. The nitrogen should be applied in 3–4 splits.
- Foliar spray with the mixture of copper oxychloride (500 g) and streptocycline or agrimycin (7.5 g) in 500 litres of water for one ha. Start spraying after 30 days of transplanting and repeat after 15 days if high disease pressure is expected.
- Practicing field sanitation such as removing weed hosts, infected plants and crop debris, rice straws, and volunteer seedlings .
- Maintaining shallow water in nursery beds, providing good drainage during severe flooding.
- Movement of irrigation water from affected field to the adjacent rice field should be restricted.
- Rouge out the infected plants from the field very carefully.

Bacterial Leaf Streak

Diagnostic Symptoms

- Small, first water-soaked, dark-green streaks on interveins from the tillering to the booting stage.
- The veins restrict the streaks' longitudinal growth, which quickly turns yellow or orange-brown.
- Bacterial exudates appeared as minute yellow or amber colored droplets along the entire length of the streaks.
- The entire leaf surface may be covered with these streaks when they combine to form enormous patches.

- When the disease is severe, the lesions turn brown to grayish white and then dry.
- An infection in the florets and seeds causes the glumes to turn brown, the ovary, stamens, and endosperm to die, and the florets and seeds to turn brown or black(Fig.-6).

Etiology

Scientific Classification

Kingdom : Prokaryotae

Division : Gracilicutes

Class : Proteobacteria

Family : Psudomonadaceae

Genus : *Xanthomonas*

Species : *campestris pv. oryzae*

- Causal organism : *Xanthomonas oryzae pv. oryzicola*
- Rod-shaped bacteria
- The bacteria don't have capsules or spores.
- They only have one polar flagellum to help them move.
- At 28 °C, they can grow well and are aerobic, gram-negative bacteria.
- The bacterial colonies on nutrient agar are round, smooth, convex, viscid, and pale yellow with an entire edge.

Disease Cycle

- Bacteria survive from season to season in infested seed andcrop debris.
- Transmission occurs by seed in summer crops, and in irrigation water.
- Bacteria enter leaves through stomata and surface damage, often caused by insects.
- Masses of bacteria develop in the parenchyma.

Epidemiology

- High humidity and a warm temperature.
- Early planting phase, from maximal tillering to the start of panicles.

Management

- The use of resistant varieties, hot water treated seeds, appropriate plant spacing, and fertilizer treatment can all help manage the disease.
- It is imperative to maintain field sanitation.
- It is possible to reduce the initial inoculum at the start of the season by destroying volunteer seedlings, straws, and rats that remain after harvest.
- Having an effective drainage system, particularly in seedbeds, can help control the disease as well.
- Planting resistant cultivars is the best way to manage leaf streak caused by bacteria.
- Grow nurseries in secluded upland environments if possible.
- When transplanting seedlings, do not cut them.
- Apply a 20% solution of fresh cowdung water extract, lemon grass extract, or mint extract.
- Applying a 300g mixture of streptomycin sulfate and tetracycline plus 1.25 kg of copper oxychloride per hectare.

Tungro

Diagnostic Symptoms

- Plants are markedly stunted.
- Leaves show yellow to orange discoloration and interveinal chlorosis.
- Young leaves are sometimes mottled while rusty spots appear on older leaves.
- Tillering is reduced with poor root system.
- Panicles not formed in very early infection, if formed, remain small with few, deformed and chaffy grains.

Etiology

- Causal organism **:** *Rice tungro bacilliform virus* (RTBV) and *Rice tungro spherical virus* (RTSV)
- Two morphologically unrelated viruses present in phloem cells.

- Rice tungro bacilliform virus (RTBV), bacilliform capsid, circular ds DNA genome.
- Rice tungro spherical virus (RTSV) isometric capsid, ss RNA genome.

Disease Cycle

- Tungro disease viruses are transmitted from one plant to another by leafhoppers that feed on tungro-infected plants.
- Most efficient vector is the green leafhopper.
- Leafhoppers can acquire the viruses from any part of the infected plant by feeding on it, even for a short time.
- It can, then, immediately transmit the viruses to other plants within 5–7 days.
- Viruses do not remain in the leafhopper's body unless it feeds again on an infected plant and re-acquires the viruses.
- Tungro incidence depends on the availability of the virus sources and vector population.
- Other than infected rice plants in the farmer's field, other primary sources for tungro, include:stubble of previous crops, new growth from infected stubbles that had not been properly plowed under and harrowed effectively, volunteer rice, infected plants in nearby rice fields
- Seedlings raised in nurseries or seedbeds can also be infected with Tungro prior to transplanting and can be a primary source of virus.
- Transplanting seedlings from nurseries in tungro-infected areas has also shown to increase infection rates in the field, particularly, in cases where seedbed is in a tungro-endemic area or when the nursery duration is 5–6 weeks.

Management

- Field sanitation, removal of weed hosts of the virus and vectors.
- Grow disease tolerant cultivars like Pankhari203, BM66, BM68, Latisail, Ambemohar 102, Kamod 253, IR50 and Co45.
- Control the vectors in the nursery by application of Carbofuran 170 g/200m^2 10 days after sowing to control hoppers.

- Spray Neem oil 3% to control the vector in the main field 15 and 30 days after transplanting.
- Set up light traps to monitor the vector population.
- Adjust planting times to when green leafhopper are not in season or abundant, if known plow infected stubbles immediately after harvest to reduce inoculum sources and destroy the eggs and breeding sites of green leaf hopper.

Khaira

Zinc deficiency

Diagnostic Symptoms

- The leaves of affected plants show chlorosis at the base.
- Large number of small, brown or bronze spots appears on the lamina surface, which coalesce to form bigger spots and ultimately the entire leaf becomes bronze coloured and dries up.
- The growth of diseased plant remains stunted.
- Root growth is also restricted and usually the main roots turn brown.
- The finer roots are destroyed.
- The plants fails to grow further and produce no panicles in severe conditions but sometimes there is natural recovery after 45 days of transplanting.
- The disease appears in nursery but may also appear in patches after 10–15 days of transplanting.

Management

- Spray mixture of zinc sulphate (5 kg) and lime (2.5 kg) in 500 litres of waters after 10 days of sowing in the nursery followed by a second spray as above after 25 days of sowing in the nursery.
- Spray can be repeated after 10–15 days of transplanting if symptoms appear in the main field.
- Zinc sulphate (5 kg) and urea (10 kg) in 500 litres of water/ ha can also be sprayed for control of the disease.

Model Question Paper

A. Objective Type Questions

a. Choose the correct answer from the following

(1) Bengal famine is related to which disease.

a. Blast of rice
b. Brown spot of rice
c. False smut of rice
d. None of the above

(2) False smut of rice is caused due to :

a. *Ustilaginoidea virens*
b. *Helminthosporium oryzae*
c. *Tilletia indica*
d. *Clavibacter tritici*

(3) The first place in India where blast disease was initially recorded to have occurred in 1918.

a. Tamil Nadu
b. Uttar Pradesh
c. Bihar
d. None of the above

(4) The perfect stage of rice blast is ?

a. *Pyricularia oryzae*
b. *Magnaporthe grisea*
c. *Ustilaginoidea virens*
d. *Clavibacter tritici*

(5) *Magneporthe grisea*produces :

a. Perithecium
b. Apothecium
c. Cleistothecium
d. Pycnidium

(6) Mucoid and straw-colored, colonies of bacteria *Xanthomonas oryzae pv. oryzae* grow slowly and eventually turn

a. Yellow.
b. Brown
c. White
d. Blue

(7). Bcterial blight of rice is managed by seeds treatment with hot water for 30 minutes at temperature

a. 50^0C
b. 54^0C
c. 58^0C
d. 64^0C

(8) Stable bleaching powder application in the rice control which disease.

a. Blast of rice
b. Brown spot of rice
c. False smut of rice
d. Bacterial blight of rice

(9) Some isolates of *Rhizocotnia solani* cusing sheath blight of rice the mycelium produces barrel shaped cells which have been termed as

a. Chlaymydospore
b. Basidiospores
c. Pycnidiospore
d. Ascospore

(10) Uyeda and Ishiyama provided a detailed description of which disease and its causative organism.

a. Blast of rice
b. Brown spot of rice
c. False smut of rice
d. Bacterial blight of rice

(11) The brown spot disease was initially documented in India in 1919 by Sundararaman from

a. Madras
b. Patna
c. Haryana
d. Kolkatta

Answer

Sl.No	Answer	Sl.No	Answer
1	b. Brown spot of rice	6	a.Yellow
2	a.*Ustilaginoidea virens*	7	b.54^{0}C
3	a.Tamil Nadu	8	d.Bacterial blight of rice
4	*b. Magnaporthe grisea*	9	a.Chlaymydospore
5	a.Perithecium	10	d.Bacterial blight of rice
		11	a.Madras

b. Fill in the blanks with suitable words

1. False smut of rice is incited by ....
2. Brown spot was thought to be the main cause of --------------- Famine.
3. The optimum temperature for development of brown spot of rice is ---------
4. Sheath blight was first documented from country----------- in 1910.
5. Leaf sheaths close to the water's edge exhibit the first symptoms of ------------disease of rice.

6. ------------------anastomosis group of *Rhizoctonia solani* is responsible for rice sheath blight in India.
7. Bacterial blight of rice was first reported from-----------------.
8. ------------------------- from Pune provided the first description of the brown spot disease of rice in India.
9. Kresek formation takes place in ----------------------- disease of rice .
10. Oozing is the sign of --------------------- disease of rice.
11. Rice tungro bacilliform virus (RTBV) has , circular -----------genome
12. Rice tungro spherical virus (RTSV) has ---------------- genome.
13. Tungro disease viruses are transmitted from one plant to another by ---- -------------- that feed on tungro-infected plants.
14. Green leaf hopper transmit the tungro viruses to other rice plants within -------------- days.
15. Foliar spray of Zinc sulphate can be used to manage----------- disease of rice.
16. Large number of small, brown or bronze spots appears on the lamina surface of rice , which coalesce to form bigger spots and ultimately the entire leaf becomes bronze coloured and dries up is a characteristic symptom of ----------- disease

Sl.No	Answer	Sl.No	Answer
1	*Ustilaginoidea virens*	9	Bacterial blight
2	Bengal famine	10	Bacterial blight
3	25-30°C	11	ds DNA
4	Japan	12	ss RNA
5	Sheath blight	13	leafhoppers
6	AG-I-1	14	5–7
7	Japan	15	Khaira
8	Srinivasan *et al*	16	Khaira

(C) State whether the following statements are *True* or *False*

1. In India, the stage that produces sclerotia most frequently occurs during the mycelial stage of *Rhizoctonia solani.*
2. Seed treatment with streptocycline is highly effective for the control of bacterial blight of rice.

3. The pathogen *Xanthomonas oryzae pv oryzae* is gram negative bacteria
4. The *Rhizoctonia solani* causing sheath blight of rice primarily survives as sclerotia and mycelia in diseased plant debris left in the field.
5. It is advisable to limit the flow of irrigation water from the affected field to the nearby rice crop to manage bacterial blight of rice.
6. The blast disease is mainly disseminated via airborne conidia.
7. Khaira disease appears in nursery but may also appear in patches after 10–15 of transplanting of rice in main field.
8. *Helminthosporium oryzae* also known as *Drechslera oryzae*.
9. Foliar spray the crop rice with Edifenphos (0.1%) manage brown spot of rice.
10. Brown spot of rice is commonly known as fungal sesame leaf spot.
11. The fungus *M. grisea* produces Fusiform, hyaline, four-celled, and slightly curved ascospores.
12. Kresek phase of bacterial blight is observed 1-3 weeks after transplanting of rice.
13. The kresek infected tillers may be confused with the stem borer injury but the latter can be easily pulled out while it is not so with the kresek infected tillers.
14. Most efficient vector of rice tungro virus is the green leafhopper.
15. Brown or bronze spots appears on the lamina surface of rice, which coalesce to form bigger spots and ultimately the entire leaf becomes bronze coloured and dries up is a characteristic symptom of khaira disease

Answer

Sl.No	Answer	Sl.No	Answer
1	True	9	True
2	True	10	True
3	True	11	True
4	True	12	True
5	True	13	True
6	True	14	True
7	True	15	True
8	True		

B. Descriptive Questions

a. Long Answer question

1. Briefly describe the diagnostic symptoms, causal organism, disease cycle and disease management of the following plant disease

   (a) Blast of rice

   (b) Brown spot of rice

   (c) Sheath blight of rice.

2. Describe in detail the symptoms, causal organism, disease-cycle and management of bacterial blight of rice.

3. Describe in detail the symptoms, causal organism, disease-cycle and management of bacterial streak of rice.

4. Describe in detail the symptoms, causal agent, and management of khaira disease of rice.

5. Write the causal organism and major symptoms of any three of the following plant disease:

   i) Brown spot of rice

   (ii) Sheth rot of rice

   (iii) False smut of rice.

6. Describe the symptoms and disease management of any two plant diseases caused by the following pathogens:

   (i) *Pyricularia oryzae*

   (ii) *Rhizoctonia solani*

   (iii) *Helminthosporium oryzae*

7. Describe in detail about integrated management schedule of blast of rice.

8. Describe the brown spot of rice under heads, diagnostic symptoms, causal organism, disease cycle and management.

9. Describe the false smut of rice under heads, diagnostic symptoms, causal organism, disease cycle and management.

10. Write down the major viral diseses of rice? Describe any one in detail under heads, diagnostic symptoms, causal organism, mode of transmission and integrated management.

11. Distinguish between
    a. Leaf blast and Nodal blast
    b. Bengal famine and Irish famine
    c. Epidemiology of Blast and Brown leaf spot.

b. Short Answer Questions

1. Differentiate between sheath blight and sheath rot of rice.
2. How would you identify the disease in the field :
   (a) False smut of rice
   (b) Bacterial blight rice
3. Write the epidemiological condition of brown spot of rice.
4. Write short notes on integrated management of bacterial blight of rice and Etiology of false smut.
5. Describe epidemiological conditions for development of shealth blight of rice.
6. How would you identify the khaira disease of rice in field.

c Very Short Answer Questions

1. Name the pathogen or causal organism of following plant disease;
   (a) Blast of rice
   (b) Brown spot of rice
   (c) Sheath blight of rice
   (d) False smut of rice
2. Explain solar energy treatment of rice seeds.
3. Describe integrated management schedule of tungro disease of rice..
4. Differentiate between symptoms of Tungro and Grassy stunt.
5. Name the three viral diseases of rice with their causal organism.
6. How can you able to identify kresek infected tillers with the stem borer injury in rice ?
7. How can you quickly diagonise the infection of bacterial blight disease in rice ?

Fig. 1: Blast

Fig. 2: Brown spot

Fig. 3: Sheath blight

Fig. 4: False Smut

Fig. 5: Bacterial blight

Fig. 6: Leaf Streak

Plate 1: Photograph showing symptoms of major diseases of Rice

2

Wheat Crop Diseases & Management

Major Diseases	**Causal Organism**
Black or stem rust	*Puccinia graminis tritici* Pers
Brown or leaf rust	*Puccinia recondita f. sp. tritici* Eriks. & Henn
Yellow or stripe rust	*Puccinia striiformis var. striiformis* Westend
Loose smut	*Ustilago tritici* (Pers.) E. Rostr.
Karnal bunt	*Tilletia indica Mitra*
Flag smut	*Urocystis agropyri (G. Preuss)* J. Schröt.
Hill bunt or Stinking smut	*Tilletia tritici (syn. Tilletia caries)* Bjerk. Wint. and *T. laevis (syn. T. foetida)*. Kuhn
Karnal bunt	*Tilletia indica* Mitra
Powdery mildew	*Blumeria graminis (DC.)* Speer
Leaf blight	*Alternaria triticina* Prasada & Prabhu

Rust

The rust of wheat has been studied more than any other disease in the world. K.C. Mehta and colleagues conducted exceptional research on the disease in India. There are three recognized wheat rusts. These include *Puccinia graminis tritici*, which causes black stem rust; *Puccinia recondita*, which causes brown rust (leaf rust); and *P. striiformis*, which causes yellow rust. Considering their later seasonal arrival, brown and yellow rusts inflict more damages than black rust.

1. Black or Stem Rust

Diagnostic Symptoms

- Symptoms are produced on almost all aerial parts of the wheat plant but are most common on stem, leaf sheaths.
- Brown pustules of uredina appear on lower surface of the leaves, leaf sheaths and the stems, and on the spikes. which give them a rusty appearance.
- Uredina form urediniospores , which are the repeating spores.
- Uredina gradually start producing teliospores and finally transform into telia.

- Telia are black and crust like, formed abundantly on the stem.
- For this reason, the rust is also called black stem rust.
- Othe stymptoms are stunted growth of the plants, and poor tilleirng , which in severe infections, grain are not formed.(Fig. 2).

Etiology

Systemic position

Phylum	Basidiomycotina
Class	Teliomyetes
Order	Uridinales
Family	Pucciniaceae
Genus	*Puccinia*
Species	*gramins tritici*

- Pathogen *;Puccinia graminis tritici.*
- Pathogen is an obligate biotroph and has a complex life cycle featuring alternation of generations.
- Pathogen is macrocyclic, heteroecious, requiring two hosts to complete its life cycle - the wheat and barberry.
- Wheat is the primary host and barberry is the alternate host.
- There are many species in *Berberis* and *Mahonia* that are susceptible to stem rust, but the common barberry is considered to be the most important alternate host.
- *Puccinia graminis* can complete its life cycle either with or without barberry (the alternate host).
- *Puccinia graminis* is macrocyclic (exhibits all five of the spore types that are known for rust fungi).
- Characteristic rust color on stems and leaves is typical of a general stem rust as well as any variation of this type of fungus.

Disease Cycle

On Wheat

- Due to its cyclical nature, there is no true ‘start point’ for this process. Here, the production of urediniospores is arbitrarily chosen as a start point.

- Urediniospores are formed in structures called uredinia, which are produced by fungal mycelia on the cereal host 1–2 weeks after infection.
- Urediniospores are dikaryotic (contain two un-fused, haploid nuclei in one cell) and are formed on individual stalks within the uredinium. They are spiny and brick-red.
- Urediniospores are the only type of spores in the rust fungus life cycle which are capable of infecting the host on which they are produced and this is therefore referred to as the 'repeating stage' of the life cycle.
- It is the spread of urediniospores which allows infection to spread from one cereal plant to another. This phase can rapidly spread the infection over a wide area.
- Towards the end of the cereal host's growing season, the mycelia produce structures called telia.
- Telia produce a type of spore called teliospores.These black, thick-walled spores are dikaryotic.
- Teliospores are the only form in which *Puccinia graminis* is able to overwinter independently of a host.
- Each teliospore undergoes karyogamy (fusion of nuclei) and meiosis to form four haploid spores called basidiospores. This is an important source of genetic recombination in the life cycle.
- Basidiospores are thin-walled and colourless.

Barberry

- Basidiospores cannot infect the cereal host, but can infect the alternative host (Usually barberry).They are usually carried to the alternative host by wind.
- Once basidiospores arrive on a leaf of the alternative host, they germinate to produce a haploid mycelium which directly penetrates the epidermis and colonises the leaf.
- Once inside the leaf the mycelium produces specialised infection structures called pycnia.
- Pycnia produce two types of haploid gametes, the pycniospores and the receptive hyphae.
- Pycniospores are produced in a sticky honeydew which attracts insects. The insects carry pycniospores from one leaf to another.

- Splashing raindrops can also spread pycniospores.
- A pycniospore can fertilise a receptive hypha of the opposite mating type, leading to the production of a dikaryotic mycelium.
- This is the sexual stage of the life cycle and cross-fertilisation provides an important source of genetic recombination.
- This dikaryotic mycelium then forms structures called aecia, which produced a type of dikaryotic spores called aeciospores.
- These have a worty appearance and are formed in chains - unlike the urediniospores which are spiny and are produced on individual stalks.
- Chains of aeciospores are surrounded by a bell-like enclosure of fungal cells.
- Aeciospores are able to germinate on the cereal host but not on the alternative host (they are produced on the alternative host, which is usually barberry).
- They are carried by wind to the cereal host where they germinate and the germ tubes penetrate into the plant.
- Fungus grows inside the plant as a dikaryotic_mycelium. Within 1–2 weeks the mycelium produces uredinia and the cycle is complete.

Disease cycle without barberry

- Since the urediniospores are produced on the cereal host and can infect the cereal host.
- It is possible for the infection to pass from one year's crop to the next without infecting the alternative host (barberry).For instance, infected volunteer wheat plants can serve as a bridge from one growing season to another.
- In other cases the fungus passes between winter wheat and spring wheat, meaning that it has a cereal host all year round.
- Since the urediniospores are wind dispersed, this can occur over large distances.
- This cycle consists simply of vegetative propagation - urediniospores infect one wheat plant, leading to the production of more urediniospores which then infect other wheat plants.

Epidemiology

- Optimum temperatures above - 20^0C.
- Availability of free water.
- Deposition of dew on leaf surface.

Management

- Use of resistant varieties like HP2278, HW741, WL614, Sonara 63,64 etc. Mostly ,the resistant cultivar used in India contain the Sr31 gene, which, unfortunately is susceptible to the fast spreading UG-99 race of the pathogen. The race originally came up in Uganda in 1999, and is fast spreading.It may soon enter South Asia, and knock at our door. The wheat breeders have been altered by GRI (Global Rust Initiative) regarding the need for developing cultivars resistant to the UG-99.
- Mixed cropping of wheat and barley with suitable crop.
- Reduction in proportion of nitrogen in the N.P.K in a fertilizer.
- Seed treatment with Carboxin or Oxycarboxin @ 2.5 g/kg seed.
- Foliar spray with Propioconazole @ 0.1 % or Pyraclostrobin 133g/ L+Epoxiconazole 50g/L @ 0.1%. if required repeat after 15days.

Annual Recurrence of Rust in India

- *Puccinia graminis tritici* is a heterocious, macrocyclic rust.
- Hoewever, in India and many other countries the alternate host, i.e. barberry , donot play any role in the recurrence of rust.
- Late Prof. K.C. Mehta has been the pioneer worker in connection with annual recurrence of rust in India.
- According to him, only urediniospores are accountable for the initiation and spread of rust from year to year.
- *Barberis* does not play any active role as an alternate host.
- The aecial stages present on barberry in India actually belong to *Aecidium montanum* and not to *Puccinia graminis tritici*.
- Teliospores do not survive at temperature over 26^0C.
- And in the Indian plains, the teliospores on wheat are produced in February-March. After March the day temperature in plains is always more than 26^0C.

- Therefore the urediniospores are the chief source of the rust infection on wheat in India.
- But urediniospores also cannot survive the high summer temperature of Indian plains.
- For that Mehta investigated and proved that these are those urediniospores that over summer in the northern hills at higher altitudes of 1300-2500 m where they survive on self sown wheat plants and tillers.
- The conditions of high altitudes and low temperature on hills are favorable for the survival of urediniospores. Hence they retain their vitality in hills.
- These surviving urediniospores of hills first infect the wheat crop near or at the foot hills, where they are carried easily by the air currents.
- From these infected wheat plants in the foot hills they are carried by the winter winds to the plains, where they infect the wheat crop in January-February.
- Mehta therefore suggested that the rust severity in the plains can be reduced if there is no wheat cultivation on the hills for some time.

2. Brown or Leaf Rust

- This is the most cmmon rust in North and South India.
- It appears earlier than the black rust.

Diagnostic Symptoms

- Most common site for symptoms is on upper leaf blades, however, sheaths, glumes and awns may occasionally become infected and exhibit symptoms.
- Symptoms begin as small, circular to oval yellow spots on infected tissue of the upper leaf surface.
- As the disease progresses, the spots develop into orange colored pustules which may be surrounded by a yellow halo .
- The pustules produce a large number of spores that are easily dislodged from the pustule resulting in an "orange dust" on the leaf surface or on clothes, hands and equipment.
- As the disease progresses, black spores may be produced resulting in a mixture of orange and black lesions on the same leaf.

- Tiny orange lesions may be present on seed heads, but these lesions do not develop into erumpent pustules(Fig. 3).
- Yield loss often occurs

Etiology

Systemic position

Kingdom : Fungi

Phylum :Basidiomyotina

Class :Teliomycetes

Order :Uredianles

Family :Pucciniaceae

Genus :*Puccinia*

Species :*recondita*

- Pathogen *Puccinia triticina (P. recondita).*
- Pathogen is a biotroph and heteroecious.
- Primary host is wheat and the secondary host is *Thalictrum.*Leaf rust is spreads via airborne spores.
- Five types of spores are formed in the life cycle.
- Uredospores, teleutospores, and basidiospores develop on wheat plants and pycnidiospores and aeciospores develop on the alternate hosts.
- Uredospores are brown, round or oblong, echinulate, stalked and have 7-10 germ pores.These spores are dispersed in the air and cause infection to other plants.
- Teliospores are bicelled with flattened top of the upper cell, brown and smooth.They are produced by teliosori, which are small, oval to linear, black in color, and covered by epidermis.
- Teliospores germinate to produce basidia and basidiospores, which normally cause infection in alternate host.
- Functional alternate host, which has not been reported from India, is *Isopyrum fumarioides.*
- In India, the non functional host is *Thalictrum,* which commonly occurs in hills.

- Spermogonial and aecial stages of the pathogen have not been reported.

Disease Cycle

- Pathogen over-summers in low and mid altitudes of Himalayas and Nilgiris.
- Wind blown urediniospores from hills, cause th primary infection.
- They germinate and penetrate through closed stomata on the lower surface, and form dikaryotic mycelium.
- Uredinia are formed and urediniospores cause secondary infections, till end of the season when teliospores are formed.
- The pycnial and aecial stages are not reported in India.

Epidemiology

- Relative humidity- 100% .
- Optimum temperature - 20–25 °C.
- Free moisture.

Management

- Collect and destroy crop debris.
- Provide irrigation at critical stages of the crop.
- Avoid water logging.
- Avoid water stress during flowering stage.
- Balance application of NPK.
- Follow mixed cropping and crop rotation
- Avoid excess application of "N".
- Foliar spray with Propiconazole 25% EC @ 200 ml in 200 l of water/acre or Tebuconazole 25% EC @ 200 ml in 200 l of water/acre or Zineb 75% WP @ 6-8 Kg in 300-400 l of water/acre or Mancozeb 75% WP @ 6-8 Kg in 300 l of water/acre.

3. Yellow or Stripe Rust

- Yellow rust does considerable damage in Punjab, Haryana and Uttar Pradesh, which are the main wheat growing areas.

- It is rare in South India because of the high heat during the crop season.

Diagnostic Symptoms

- Mainly occur on leaves than the leaf sheaths and stem.
- Bright yellow pustules (Uredia) appear on leaves at early stage of crop and pustules are arranged in linear rows as stripes.
- Stripes are yellow to orange yellow and this gives the name 'yellow' rust to the disease.
- In severe infections, all parts of the plant are infected, and the stripes disappear due to the crowding of uredinia.
- Teliospores are also arranged in long stripes, dull black in colour.
- Telia appear at the end of the season, as black streaks, on the lower surface(Fig. 4).

Etiology

Systemic position

Phylum	Basidiomycotina
Class	Teliomyetes
Order	Uredinales
Family	Pucciniaceae
Genus	*Puccinia*
Species	*striiformis*

- Pathogen ;*Puccinia striiformis*
- Pathogen is biotroph and its primary host is wheat.
- Uredospores of rust pathogen are almost round or oval in shape and bright orange in colour.
- Each uredospores possesses 6-10 scattered germ pores.
- Teliospores are bright organge to dark brown, two celled and flattened at the top.
- Sterile paraphyses are also present at the end of sorus.
- No intervening hosts for pycnial and aecial stages of the pathogen have been discovered.

Disease Cycle

- Inoculum survives in the form of uredospores /teliospores in the hills during off season on self sown crop or volunteer hosts, which provide anexcellent source of inoculum.
- About ninety weeds serve as collateral hosts, which can provide the urediniospores for primary infection.
- Wind borne uredospores from hills are lifted due to cyclonic winds cause primary infection in the plains during crop season.
- In India, role of alternate host (Barberis) is not there in completing the life cycle.
- Urediniospores spread the disease until the end of the season, when telia are formed.
- Teliospores have no role in disease causation.

Epidemiology

- Low temperature -10-15°C.
- Relative humidity- >60 percent.

Management

- Grow resistant varieties like PBW 343, PBW 550, PBW 17.
- Collect and destroy crop debris.
- Provide irrigation at critical stages of the crop.
- Avoid water logging.
- Avoid water stress during flowering stage.
- Balance application of NPK.
- Follow mixed cropping and crop rotation.
- Avoid excess application of "N".
- Foliar spray with Propiconazole 25% EC @ 200 ml in 200 l of water/ acre or Tebuconazole 25% EC @ 200 ml in 200 l of water/acre or Zineb 75% WP @ 6-8 Kg in 300-400 l of water/acre or Mancozeb 75% WP @ 6-8 Kg in 300 l of water/acre.

A comparison of the three wheat rusts

	Leaf rust	**Stem rust**	**Stripe rust**
Pathogen	*Puccinia triticina (P. recondita)*	*Puccinia graminis tritici*	*Puccinia striiformis*
Month of appearance	January	March-April	December-January
Pustule location	Leaf, mainly on the upper surface	Stem and leaf, upper and lower surfaces of leaf; occasionally on head and seeds	Leaf, upper surface; occasionally on head and seeds
Pustule color	Orange-brown	Orange-red to dark-red	Orange-yellow
Pustule arrangement	Single and random	Single and random	Stripes
Pustule shape and size	Round or slightly elongated; small to medium	Oval shaped or elongated; small to large	Round, blister-like; small
Tearing of host epidermis	Rare	Conspicuous	None
Uredospores	Spherical, brown, echinulate, wall possesses 7-10 germ pores, 16-20 mm in diameter	Oval, echinulate, brick red or rusty in appearance, posses thick cellulose wall containing four germ pores situated in an equatorial , 25-30x17-20µm	Spherical to ovate, wall colorless and finally echinulate possessing 6-16 germ pores, vary variable in size from 23-35 by 20-35µm
Teliospores	Resemble those of yellow rust in shape and size, top cell flattened at apex, sorus divided into small chambers by paraphyses	Thick walled, smooth, bicelled, super-imposed cells, the top cell being rounded or thickened at the apex, dark brown and measure 40-60x15-20µm	Dark brown, bicelled, the top cell flattened at the apex, in contact with the epidermis, interspersed with brown and unicellular paraphyses measure 35-63x12-24µm
Optimum temperature for infection	15-20°C	15-29°C	7-12°C
Optimum temperature for disease development	20-25°C	26-30°C	10-15°C
Alternate hosts	Meadow rue (*Thalictrum* sp.)	Barberry	Not known
Annual recurrence	By uredospores from hills	By uredospores from hills	By uredospores from hills
Survival altitudes	About 1450 -2050 meter	About 1500 meter	About 1900 meter

4 Loose Smut

Diagnostic Symptoms

- Infected plants is very difficult to detect in the field until heading.
- Infected heads emerge earlier than normal heads.
- Entire inflorescence is commonly affected and appears as a mass of olive-black spores, initially covered by a thin gray membrane.
- Once themembrane ruptures, the head appears powdery.
- Spores are dislodged, leaving only the rachis intact.
- In some cases remnants of glumes and awns may be present on the exposed rachis.
- Smutted heads are shorter than healthy heads due to a reduction in the length of the rachis and peduncle.
- While infected heads are shorter, the rest of the plantis slightly taller than healthy plants(Fig. 1).

Etiology

Systemic position

Kingdom Fungi

Division Eumycota

Phylum	Basidiomycotina
Class	Teliomyetes
Order	Ustilaginales
Family	Tilletiaceae
Genus	*Ustilago*
Species	*tritici*

- Pathogen: *Ustilago nuda tritici*
- Mycelium is hyaline primarily but turns to brown near maturity.It is septate, dikaryotic, and grows systemically inside the plant.
- Mycelia cells get transformed into brown, spherical, echinulate teleutospores.
- Teleutospores measure 5-9 μm in diameter, germinate readily producing promycelium or basidium consisting of four uninucleate cells.

- Basidium, unlike other smut causing fungi, do not produces basidiospores.
- Its cells germinates and give rise to uninucleate hyphae, the primary hyphae; a pair of sexually compatible uninucleate primary hyphae fuse and results in dikaryotic hypha often called infection hypha.

Disease Cycle

- The pathogen perennates as dormant mycelium lying inside the kernel (seed), which has been infected by the infection hypha.
- Mycelium lies dormant within the seed until next season when they are sown.
- On seed germination and growth of the seedling, the mycelium becomes active, starts developing , and moves systemically along the growing seedling.
- Mycelium is hyaline during its growth through the plant, but it turns to brown at maturity.
- When the head emerges, the hyphae accumulate in the floral parts.Their cells are transformed into teleutospores, which completely fill in the spikelets.
- They first remain covered by adelicate silvery membrane, which soon bursts spreading almost all teleutospores to atmosphere.
- They are now wind blown leaving a bare rachis behind. This is usually the time when the healthy heads of adjacent plants are in flowering stage.
- Disseminated teleutospores lodge between the glumes and reach the feathery stigma germinate thereupon in moist stigmatic fluid giving out promycelium or basidium consisting of four cells.
- The basidium produces no basidiospores, but its cells germinate and produce short uninucleate, sexually compatible primary hyphae that fuse in pairs and give rise to dikaryotic mycelium the infection thread.
- Infection hypha penetrates the flower through the stigma or the young ovary wall and become established in the pericarp, integuments, and in the tissues of the embryo before the kernels become mature.
- Mycelium then becomes inactive and remains dormant, primarily in scutellum of the kernel.
- Fungal presence in no way happers the grain formation .

- When such infected kernels are sown in the next season the hyphae become active and show further course of their action.

Epidemiology

- Frequent rain showers.
- High humidity.
- Cool temperatures (16-22^{0}C) during flowering season.

Management

- Use of resistant varieties like NP710, 718,761,770, Bansipali 808, Bansi 224 etc.
- **Hot water treatment**: The seeds are soaked in water for 4-6 hours at 68-86^{0}F and then placed in water for 2 minutes at 120^{0}F, and finally taken out, dried and sown.
- **Solar energy treatment**: The seeds are soaked into water for four hours (8 am-12 noon), then taken out and dried in sun for four hours (12 noon-4 pm), and then sown.
- Treat the seed with Vitavax @ 2g/kg seed or Benomyl 50 % WP @ 2g/ Kg seeds or Carbendazim 50% WP @ 2 g/ Kg seeds or Carboxin 75% WP @ 2 -2.5 g/Kg seeds or Tebuconazole 2% DS @ 0.2 Kg/10 Kg seed or Carboxin 37.5% + Thiram 37.5% DS @ 3.0 g/Kg before sowing.
- In the standing crop, the plants showing yellowing of the boot leaf tip normally are the ones which will give smutted ear heads on emergence. Uproot such plants before ear emergence to reduce the infestation of healthy seeds at later stage.
- Use disease free seeds in the healthy field. For seed production, disease free field/areas to be identified for having crop without considerable inoculum load.
- Burry the infected ear heads in the soil, so that secondary spread is avoided.

5. Karnal Bunt

Diagnostic Symptoms

- Symptoms of karnal bunt are often difficult to distinguish in the field owing to the fact that incidence of infected kernels on a given head is low.

- Symptoms are most readily detected on seed after harvest.
- Black sorus, containing dusty spores is evident on part of the seed, commonly occurring along the groove.
- Heavily infected seed is fragile and the pericarp ruptures easily.
- Foul, fishy odor associated with karnal bunt.
- Odor is causedby the production of trimethylamine by the pathogen.
- Seed that is not extensively infected maygerminate and produce healthy plants(Fig. 5).

Etiology

Scientific Classification

Kingdom : Fungi

Division: Eumoycota

Sub-Division : Basidiomyotina

Class : Teliomycetes

Order : Ustilaginales

Family : Tilletiaceae

Genus : *Neovosia*

Species : *indica*

- Pathogen: *Neovosia indica.*
- Telitospores: Spherical to oval with reticulation looking as curved spines on the episore, dark brown and measure 20-49 μ in diameter.
- Telitospore wall consists of three layers; perisporium, episporium and endosporium.
- On germination, a teleutospore gives rise to a promycelium that bears long, sickle shaped sporidia in a cluster at its tip; each cluster or whorl generally consists of 60-185 sporidia called primary sporidia and are unicellular.
- Primary sporidia are incompatible with each other, so produce another crop of sporidia called secondary sporidia.

- The compatible primary sporidia and secondary sporidia fuse and develop into the dikaryotic hyphae or sporidia.
- The infection is caused by these dikaryotic hyphae or sporidia.

Disease Cycle

- Disease is soil borne and pathogen perennates through teleutospores, which reach the soil either directly falling on the ground or through sticking on the surface of infected seeds and survive until flowering time of the next wheat crop.
- When favourable conditions become available at the time of flowering of the wheat crop, the perennating teleutospores germinate producing short, stout promycelia which give rise to primary sporidia in whorl.
- Primary sporidia produce secondary sporidia.
- Sporidia are wind or water splash disseminated to the flowers whereupon the compatible primary and secondary sporidia fuse producing dikaryotic hyphae and or sporidia.
- These are the dikaryotic hyphae and or sporidia which cause primary infection of individual florets.
- Secondary spread of the pathogen within and between spikelets takes place through secondary sporidia and dikaryotic mycelia produced on the primary infected spikelets.

Epidemiology

- Relative humidity over 70% favors teliospore development.
- Day time temperatures - 18-24 °C.
- Soil temperatures -17-21 °C.

Management

- Grow resistant / tolerant varieties.
- Low-lying areas of the field accumulate water and are more prone to kernal bunt. Effective land leveling and drainage can reduce disease incidence.
- Decrease seed rate during sowing.
- Increase row spacing during sowing.

- Delayed sowing.
- Avoiding irrigation during the period of awn emergence and end of flowering may hinder disease development.
- Avoid lodging by using balance dose of Nitrogen and Potash.
- Plastic mulching or solarization reduces the chance of teliospore germination.
- Burn the stubble after harvesting.
- Crop rotation with non-host crop.
- Foliar spray with Propiconazole 25% EC @ 200 ml in 300 l of water/acre or Bitertanol 25% WP @ 896 g in 300 l of water/acre or Thiram 75% WS @ 10-12 g in 400 ml of water/acre.

Model Question Paper

A. Objective Type Questions

a. Choose the correct answer from the following

(1) Which of the following is an internally seed borne disease.

a. Powdery mildew of wheat　b. Brown or leaf rust of wheat

c. Loose smut of wheat　d. None of the above

(2) Karnal bunt of wheat is caused due to :

a. *Puccinia graminis tritici*　b. *Erysiphe graminis tritici*

c. *Tilletia indica*　d. *Clavibacter tritici*

(3) Which of the following rust appears earliest on wheat ?

a. Brown or Leaf rust of wheat　b. Black or Stem rust of wheat

c. Yellow or Stripe rust of Wheat　d. None of the above

(4) Who studies the detailed disease cycle of cereal rusts in India?

a. R.S. Singh　b. K.C. Mehta

c. B.B. Mundkur　d. E.J. Butler

(5) *Clavibacter tritici* is the causal organism of :

a. Flag smut of wheat　b. Black or Stem rust of wheat

c. Tundu or Yellow ear rot of wheat　d. Leaf blight of wheat

(6) *Neovosia indica* is a soil borne pathogen perennates through

a. Teleutospores
b. Uredeospores
c. Basidiospore
d. Aeciospore

(7) The *Neovosia indica* belongs to which class.

a. Teleiomycetes
b. Oomycetes
c. Plectomycetes
d. Bsidiomycetes

(8) Foul, fishy odor associated with which disease of wheat.

a. Powdery mildew
b. Brown leaf rust
c. Loose smut
d. Karnal bunt

(9) Foul, fishy odor associated with karnal bunt disease of wheat due to the production of chemical.

a. Trimethylamine
b. Uric acid
c. Hydrocynic acid
d. Succinic acid

(10) The mycelia cells of *Ustilago tritici* get transformed into brown, spherical, echinulate

a. Teleutospores
b. Uredeospores
c. Basidiospore
d. Aeciospore

(11) According to K.C.Mehta which spore are accountable for the initiation and spread of wheat rust in India from year to year.

a. Teleutospores
b. Uredeospore
c. Basidiospore
d. Aeciospore

(12) Teliospores of *Puccinia graminis tritici* do not survive at temperature over

a. 26^0C.
b. 15^0C.
c. 20^0C
d. 10^0C.

(13) Teliospores of *Puccinia recondita* germinate to produce basidiospores, which normally cause infection

a. Alternate host.
b. Main host
c. Both
d. None

(14) The *Puccinia recondita* spores that formed on wheat plant

a. Uredospores | b. Teleutospores
b. Basidiospores | d. All

(15) The spores produced by *Puccinia graminis tritici* on alternate host

a Pycnidiospore and Aeciospore
b Pycnidiospore and Basidiospore
c Pycnidiospore and Teliospore
d Uredeospore and Aeciospore

(16) UG-99 is a race of the

a. *Puccinia graminis tritici* | b. *Erysiphe graminis tritici*
c. *Tilletia indica* | d. *Clavibacter tritici*

(17) The race of *Puccinia graminis tritici* originally came up first time in

a. Uganda | b. USA
c. Ethiopia | d. China

Answer

Sl.No	Answer	S. No	Answer
1	c. Loose smut of wheat	9	a. Trimethylamine
2	c. *Tilletia indica*	10	a.Teleutospores
3	a. Brown or Leaf rust of wheat	11	b.Uredeospore
4	*b.* K.C. Mehta	12	a.26^{0}C
5	c.*Tundu* or Yellow ear rot of wheat	13	a. Alternate host
6	a.Teleutospores	14	d. All
7	a.Teleiomycetes	15	a.Pycnidiospore and Aeciospore
8	d.Karnal bunt	16	a.*Puccinia graminis tritici*
		17	a. Uganda

b. Fill in the blanks with suitable word (s)

1. Loose smut of wheat is incited by …………..
2. *Ustilago tritici* causes ………………. disease.
3. Of the three rusts in wheat ………… requires lowest temperature for disease developement.

4. .... is known for his outstanding contribution on cereal rusts in India.

5. *Clavibacter tritici* is the causal organism of ....

6. Smuts are obligate parasite, except ----------------------, which grows well in the soil.

7. The term bunt is a distortion of --------------, while smut is a German word --------------

8. Smut diseases caused by Tilletia are called---------------------

9. The fishy stink emitted in stinking smut of wheat is due to the chemical-----------produced bt the fungus.

10. Loose smut of wheat and loose smut of --------------------are same except in host specificity.

Sl.No	Answer	Sl.No	Answer
1	*Ustilago tritici*	6	
2	Loose smut of wheat	7	
3	50-59°F	8	
4	K.C. Mehta	9	HCN
5	Tundu or Yellow ear rot of wheat	10	Barley

c. State whether the following statements are *True* or *False*

(1) Disease cycle of wheat rusts in India was described by K.C. Mehta.

(2) Seed treatment with Captan is highly effective for the control of loose smut of wheat.

(3) The rotten fish smell produced in the karnal bunt infected wheat fields is due to the production of trimethylamine by the causal fungus.

(4) *Puccinia graminis tritici* is an autoecious rust fungus.

(5) Powdery mildew of wheatis caused by *Erysiphe graminis tritici.*

(6) The *Neovosia indica* infection in wheat is caused by dikaryotic hyphae or sporidia.

(7) The teleutospore of *Neovosia indica* on germination gives rise to a promycelium that bears long sickle shaped sporidia in wheat crop.

(8) The symptoms of karnal bunt are most readily detected on wheat seed after harvest.

(9) The symptoms of karnal bunt are often difficult to distinguish in the wheat field owing to the fact that incidence of infected kernels on a given head is low.

(10) Loose smut infected plants is very difficult to detect in the wheat field until heading.

(11) Loose smut infected heads are shorter, the rest of the wheat plant is slightly taller than healthy plants.

(12) Urediniospores are the chief source of the rust infection on wheat in India.

(13) Barberis does not play any active role as an alternate host of black rust of wheat in India

S. No	Answer	S. No	Answer
1	True	8	True
2	True	9	True
3	True	10	True
4	True	11	True
5	True	12	True
6	True	13	True
7	True		

B. Descriptive Questions

a. Long Answer Questions

1. Briefly describe the diagnostic symptoms, causal organism, disease cycle and disease management of the following plant disease

   (a) Brown or Leaf rust of wheat

   (b) Loose smut of wheat

   (c) Powdery mildew of wheat.

2. Describe in detail the symptoms, causal organism, disease-cycle and management of Karnal bunt of wheat.

3. Write the causal organism and major symptoms of following plant disease:

   (i) Yellow or Stripe rust of wheat

   (ii) Seed gall of wheat

   (iii) Anguina leaf blight of wheat.

5. Describe the symptoms and disease management of any two plant diseases caused by the following pathogens:

   (i) *Ustilago tritici*

   (ii) *Puccinia graminis tritici*

   (iii) *Anguina tritici*

6. How do the three rusts of wheat differ? Describe in brief the different stages of any heteroecious rust occurring on wheat.

7. Describe the present views or recent work on annual recurrenc of wheat rusts of India.

8. Describe the loose smut of wheat under heads, diagnostic symptoms, causal organism, disease cycle and disease management.

9. Which are the important bunt diseases of wheat? Describe any one in details..

10. Distinguish between

    (i) Localized and Systemic smut infection

    (ii) Monocyclic and polycyclic smut diseases

    (iii) *Tilletia caries* and *Tilletia foetida*

11. Write short notes on

    (i) Life cycle of loose smut

    (ii) Management of smut disease

    (iii) Loose smut of wheat

b. Short Answer Questions

1. Differentiate between yellow rust and brown rust of wheat.

2. How would you identify the flowing disease in the field

   (a) Loose smut of wheat

   (b) Black or Stem rust of wheat

3. Give the disease cycle of powdery mildew of wheat.

c. Very short answer questions

1. Name the pathogen or causal organism of following plant disease;

    (a) Ear-cockle disease of wheat

    (b) Yellow or Stripe rust of wheat

    (c) Yellow ear rot of wheat

    (d) Powdery mildew of wheat.

2. Explain solar energy treatment of wheat seeds.
3. Give the different spore stages of any heteroecious rust.
4. Differentiate between smut and rust.
5. Name the three rust diseases of wheat with their causal organism.

Fig. 1: Loose smut

Fig. 2: Stem rust

Fig. 3: Brown rust

Fig. 4: Yellow rust

Fig. 5: Karnal Bunt

Plate 2: Photograph showing symptoms of major diseases of wheat

3

Maize Crop Diseases & Management

Major Diseases	**Causal Organism**
Banded leaf and sheth blight	*Rhizoctonia solani*
Southern leaf blight	*Helminthosporium maydis*
Northern leaf blight	*Exserohilum turcicum*
Downy mildew	*Peronosclerospora sorghi*
Bacterial stalk rot	*Erwinia dissolvens*
Rust	*Puccinia sorghi*
Head smut	*Sphacelotheca reiliana*
Charcoal rot	*Macrophomina phaseolina*
Rust	*Puccinia sorghi*
Common smut	*Ustilago zeae* or *Ustilago maydis*

Banded Leaf and Sheath Blight

Diagnostic Symptoms

- From the crop's seedling stage until crop maturity, this pathogen can infect the plant at any point in its growth.
- The disease initially appears on the first and second leaf sheaths above the ground, and it eventually spreads to the ears, where it results in ear rot.
- The hallmark symptoms of ear rot include the presence of light brown, cottony mycelium on the plant's ear, small, rounded black sclerotia, early ear drying, and caking of the ear sheath.
- When there is humidity, the infection that first appears on the lower sheath extends to the upper sheaths, causing leaf sheath rot and complete leaf drying.
- When the condition worsens, the majority of the plant's leaves become blighted, and easily detachable sclerotia form on the lesions.
- The disease manifests as a straw-colored, wet, irregular to roundish patch on both leaf surfaces on the basal sheath.

- Mycelium is growing profusely in the space between the stem and the leaf sheath as well as on the leaf sheath affected area.
- Eventually, as the disease worsens, many sclerotial bodies are discovered growing on the plant's afflicted areas.
- All of the maize plant's aerial sections, with the exception of the tassel, exhibit disease symptoms.
- Leaf and sheath blight is one way that the disease shows up on leaves, stalks, ears, and leaf sheaths (Fig.-1).
- A disease that affects plants 30 to 40 days old during the pre-flowering period of the plant causes severe blights and the apical part of growing maize plants to die.

Etiology

Classification

Sub division – Deuteromycotina

Class - Deuteromycetes

Sub class - Hyphomycetidae

Order - Agonomycetales

Genus - *Rhizoctonia*

Species - *solani* Kuhn

- Anamorph: *Rhizoctonia solani.*
- Teleomorph: *Thanatephorus cucumeris.*
- *Rhizoctonia solani* immature colonies appear almost white in the media, but as the colonies age, they take on a hint of brown.
- The consistent features that set this fungus apart are its many colors of brown.
- The majority of hyphae have a diameter that ranges from 5 to 14 μm.
- The most characteristic feature of *R. solani* is its dolipore septa.
- Although it's not always accurate, right-angled mycelium branching is sometimes thought to be a unique characteristic of this fungus.
- Two consistent and dependable traits of *R. solani* seem to be the constriction of branch hyphae at the place of origin and the creation of septum at a right angle in the point of origin.

- *R. solani's* sclerotia are initially grey-white in color, but later on they turn to brown to black in colour.
- They are subglobose and in shape they are slightly flattened varying from 0.5-5.0 mm in size.

Disease cycle

- The primary sources of inocolum are the sclerotia found in the soil or in the infected host debris, as well as the active mycelium that is developing close to the maize plant in the field.
- Seeds are not thought of as the source of inoculum and may not have a significant function in an outbreak of serious disease.
- Sclerotia that live on plant debris emerge in the soil during land preparation and other operations, come into contact with freshly planted, healthy seedlings, and infect them.
- After 48 to 72 hours, the infection spreads upward in bands, destroying the tissue in its path and reaching the distal half of the leaf lamina, where patches of buff white mycellial growth appear. In the 30 to 40 day older plants that are in the pre-flowering stage, sclerotia become active.
- When diseased leaves or sheaths come into contact with healthy plants, the disease spreads secondary. The entire plant is affected by the disease up to the mid-dough stage, including the ear that leaves the tassel.
- Affer ear infections happen in the later stages, affecting seed germination and causing seed rot and seedling blight.
- In maize crops, there has never been evidence of a secondary disease spread by basidiospores.

Epidemiology

- The ideal temperature range for the disease to manifest is between 28 and 30°C, with a relative humidity of 90 to 100%.
- Heavy rains and high relative humidity have a significant impact on the spread of disease.
- In the first two weeks of infection, rainfall exceeding 100 mm encourages early infection and disease development in plants.

Management

- Select field with good drainage.
- Sanitation and removal of previous crop debris/wheat straw.
- Deep ploughing 2 to 3 times at 10 to 15 days interval to destroy crop debris & weeds.
- Proper seed bed preparation, planting seed in warm, fairly moist soil (above 12.8°C).
- Use of resistant/tolerant hybrids;Pratap Kanchan-2,Pratap Makka -3,Shaktiman -1, Shaktiman -3.
- Adoption of crop rotation.
- Use balanced soil fertilizer, avoid high level of N and low level of K .
- Stripping of lower leaves along with their sheath.
- Seed treatment with *Trichoderma harzianum* 2.0% WP @ 20 g/kg of seeds.
- Foliar spray Azoxystrobin 18.2% + Difenconazole 11.4% w/w SC (Amistar Top 325 SC) 1ml/L of water at 50 DAS or immediately after symptoms appearance. If needed, repeat the spray at 15 days interval.

Southern Leaf Blight

Diagnostic Symptoms

- Disease affects leaves, leaf sheaths, ear, and maize grains.
- **On leaves**, the grayish-red or stramineous long lesions with dark brown center appear along leaf veins, being spindle-shaped or elliptical.
- Length of the lesions is 40 mm, and their width is about 6 mm.
- Lesions can coalesce, causing death of leaves.
- **On sheaths,** the lesions are brown with purple border.
- Length of the lesions is 50 mm.
- **On ears,** the lesions are spindle-shaped, brown with dark border.
- Germ part of seeds becomes dark, and the seeds lose germinating ability (Fig.-2).

Etiology

Scientific Classification

Kingdom : Fungi

Division : Eumycota

Sub- division : Deuteromycotina

Class : Hyphomycetes

Order : Hyphomycetales (Moniliales)

Family : Dematiaceae

Genus : *Helminthosporium*

Species : *maydis*

- Pathogen: *Helminthosporium maydis* (Syn: *H. turcicum*).
- Facultative parasite.
- Teleomorph stage is *Cochliobolus heterostrophus* (Drechsler) Drechsler.
- Produces a specific T-toxin causing the same symptoms of disease as those caused by conidia.
- Conidiophores are in group, geniculate, mid dark brown, pale near the apex and smooth.
- Conidia are fusiform, markedly curved, pale to mid dark golden brown with 5-11 septa.
- Size of conidia is 25-115 x 8.5-20.6 microns.
- Race T appeared on maize hybrids with Texas male sterile cytoplasm.
- Disease development was related to the growing of maize with T-cytoplasm on large areas.
- Since 1990, the use of cultivars with T male sterile cytoplasm has been forbidden in selection.
- Presently, the cultivars with M and C types of sterility are used.

Disease Cycle

- Seed-borne.
- Infected seeds and corn residues are sources of the primary infection.

- Secondary infection through wind borne conidia.
- Wind spreads the spores long distance.
- Also infects sorghum, wheat, barley, oats, sugarcane etc.

Epidemiology

- Optimum temperature - 25-31 ^{0}C.
- Relative humidity- 90-100 %.
- Presence of free water on the leaf.
- Infection takes place early in the wet season.

Management

- Use of disease-resistant hybrids of maize.
- Seed treatment with Captan or Thiram at 4 g/kg.
- Foliar spray with Mancozeb @ 0.2%, if required repeat after 15 days.
- Removal of infected maize residues.
- Crop rotation.
- Removal of crop residues that contain the infection.

Northern Leaf Blight

Diagnostic Symptoms

- Characteristic symptum is the one-to-six inch long cigar-shaped gray-green to tan-colored lesions on the lower leaves.
- As the disease develops, the lesions spread to all leafy structures, including the husks.
- Lesions may become so numerous that the leaves are eventually destroyed causing major reductions in yield due to lack of carbohydrates available to fill the grain.
- Leaves then become grayish-green and brittle, resembling leaves killed by frost.
- Yield losses can reach as high as 30-50% if the disease establishes itself before tasseling (Fig.-3).

Etiology

Scientific Classification

Kingdom : Fungi

Division : Eumycota

Sub- division : Deuteromycotina

Class : Hyphomycetes

Order : Hyphomycetales (Moniliales)

Family : Dematiaceae

Genus : *Helminthosporium.*

Species : *turcicum*

- Pathogen*: Exserohilum turcicum.*
- Mycelium is localized in the infected lesion.
- Conidiophores emerge through stomata and are simple, olivaceous, septate and geniculate.
- Conidia are olivaceous brown, 3-8 septate and thick walled

Disease Cycle

- Pathogen overwinters as mycelia and conidia on corn residues left on the soil surface.
- Conidia are transformed into thick-walled resting spores called chlamydospores.
- New conidia are produced on the old corn residue, and the conidia are carried by the wind or rain to lower leaves of young corn plants.
- Secondary spread within fields occurs by conidia produced on the leaf tissues.

Epidemiology

- Presence free water on the leaf surface.
- Temperature -18-27°C.

Management

- Planting resistant hybrids.
- Seed treatment with Captan or Thiram at 4 g/kg.
- Foliar spray with Mancozeb @ 0.2%, if required repeat after 15 days.
- A one- to two-year rotation away from corn.
- Destruction of old corn residues by tillage.

Downy mildew/Crazy top

- Sorghum downy mildew - *Peronosclerospora sorghi*
- Philippine downy mildew - *Peronosclerospora philippinensis*
- Crazy top - *Sclerophthora macrospora*

Diagnostic Symptoms

- Most characteristic symptom is the development of chlorotic stripes along almost the whole length of the leaves.
- Stripes are yellowing white in the start, continue as such, and turn darker quite late.
- Plants exhibit a stunted and bushy appearance owing to shortening of the internodes.
- White downy growth is seen on the lower surface of leaf.
- Downy growth furthermore occurs on bracts of green unopened male flowers in the tassel.
- Small to large leaves are noticed in the tassel.
- Proliferation of auxillary buds on the stalk of tassel and the cobs is common **(Crazy top)** (Fig.-4).

Etiology

Scientific Classification

Kingdom : Fungi

Division : Eumycota

Class : Oomycetes

Order : Peronosporales

Family : Peronosporaceae

Genus : *Peronosclerospora*

Species : *sorghi, philippinensis* and *sacchari*

- Hemibiotroph.
- Pathogen grows as white downy growth on both surface of the leaves, consisting of sporangiophores and sporangia.
- Sporangiophores are quite short and stout, branch profusely into series of pointed sterigmata which bear hyaline, oblong or ovoid sporangia (conidia).
- Sporangia germinate directly and infect the plants.
- In advanced stages, oospores are formed which are spherical, thick walled and deep brown.

Disease Cycle

- Primary source of infection is through oospores in soil or dormant mycelium present in the infected maize seeds or conidia produced in over wintering crop debris and infected neighboring plants.
- Oospores in the soil germinate in response to root exudates from susceptible maize seedlings.
- Germ tube infects the underground sections of maize plants leading to characteristic symptoms of systemic infection including extensive chlorosis and stunted growth.
- Whole plants show symptoms, if the pathogen is seed borne.
- Once the pathogen has colonized host tissue, sporangiophores emerge from stomata and produce sporangia which are wind and rain splash disseminated and initiate secondary infections.
- Sporangia are always produced in the night, fragile and cannot be disseminated more than a few hundred meters and do not remain viable for more than a few hours.
- Germination of sporangia is dependent on the availability of free water on the leaf surface.
- Oospores are produced in large numbers,as the crop approaches senescence.

Epidemiology

- Optimum temperature- 21-25°C.
- High relative humidity -90 per cent.
- Free water on leaf surface.
- Young plants are highly susceptible.

Management

- Deep ploughing.
- Grow resistant varieties and hybrids *viz.* CO1, COH1 and COH_2.
- Seed treatment with Metalaxyl at 6g/kg.
- Foliar spray with Metalaxyl + Mancozeb @ 0.2% on 20th day after sowing.
- Crop rotation with pulses.
- Rogue out infected plants.

Model Question Paper

A. Objective Type Questions

a. Choose the correct answer from the following

(1) Spindle shaped or elliptical lesion with size 40 mm,& width 6 mm on leves is diagnostic symptoms of disease of maize.

a. Powdery mildew b. Crazy top

c. Southern leaf blight d. Northern leaf blight

(2) Teleomorph stage of pathogen *Helminthosporium maydis* is

a. *Helminthosporium turcicum* b. *Erysiphe graminis*

c. *Tilletia indica* d. *Cochliobolus heterostrophus*

(3) *Helminthosporium maydis* belong to order

a. Hyphomycetales b. Oomycetales

c. Albuginales d. None of the Above

(4) Downy mildew of maize is caused by

a. *Helminthosporium turcicum*
b. *Erysiphe graminis*
c. *Peronosclerospora sorghi*
d. *Cochliobolus heterostrophus*

(5) Southern leaf blight disease development was related to the growing of maize with

a. T-cytoplasm
b. C-cytoplasm
c. M-cytoplasm
d. K-cytoplasm

(6) The one-to-six inch long cigar-shaped gray-green to tan-colored lesions on the lower leaves of maize is characteristic symptom of

a. Powdery mildew of maize
b. Crazy top
c. Northern leaf blight.
d. Southern leaf blight

(7) The conidia of *Helminthosporium turcicum* are transformed into thick-walled resting spores under adverse condition called

a. Chlamydospore
b. Uredospore
c. Macroconidia
d. Microconidia

(8) Proliferation of auxillary buds on the stalk of tassel and the cobs of maize infected with downy mildew is called

a. Low top
b. Crazy top
c. High top
d. Wavy top

(9) Teleomorph stage of pathogen *Rhizoctonia solani* is

a. *Helminthosporium turcicum*
b. *Thanatephorus cucumeris*
c. *Tilletia indica*
d. *Cochliobolus heterostrophus*

(10) The most characteristic feature of *R. solani* is its

a. Dolipore septa
b. Clamp connection
c. Ascospore formation
d. Basidiospore formation

S. No	Answer	S. No	Answer
1	c. Southern leaf blight	6	c. Northern leaf blight
2	d. *Cochliobolus heterostrophus*	7	a. Chlamydospores
3	a. Hyphomycetales	8	b. Crazy top
4	c. *Peronosclerospora sorghi*	9	b. *Thanatephorus cucumeris*
5	a. T-cytoplasm	10	a. *Dolipore septa*

(b) State whether the following statements are True or False

1. Race T of *Helminthosporium maydis* appeared on maize hybrids with Texas male sterile cytoplasm..
2. Foliar spray with Metalaxyl + Mancozeb @ 0.2% manage downy mildew of maize.
3. Infected seeds and corn residues are sources of the primary infection of southern leaf blight disease.
4. The pathogen *Peronosclerospora sorghi* is hemibiotroph.
5. Primary source of downy mildew infection in maize is oospores in soil or dormant mycelium present in the infected seeds .
6. Oospores are produced in large numbers in the mize infected with downy mildew, as the crop approaches senescence.
7. Young maize plants are highly susceptible to downy mildew than older plants.
8. Most characteristic symptom downy mildew of mize is the development of chlorotic stripes along almost the whole length of the leaves.
9. *Helminthosporium maydis is* facultative parasite in nature .
10. *Helminthosporium maydis* belong to family dematiaceae.
11. *R. solani's* sclerotia are initially grey-white in color, but later on they turn to brown to black in colour.
12. The primary sources of inocolum of *Rhizoctonia solani* are the sclerotia found in the soil or in the infected host debris.

Answer

S. No	Answer	S. No	Answer
1	True	7	True
2	True	8	True
3	True	9	True
4	True	10	True
5	True	11	True
6	True	12	True

B. Descriptive Questions

a. Long Answer

1. Dfferentiate between Southern and Northern blight disese of maize on the basis of the causal organisms, symptomatology, etiology, disease cycle and management.

2. Describe in detail the occurrence importance, symptoms, etiology, disease cycle and management of banded leaf and sheath blight of maize.
3. Give the diagrammatic representation of disease cycle of the following diseases of maize
   i. Downy mildew
   iii. Banded leaf and sheath blight
   iv. Powdery mildew
4. Briefly describe the diagnostic symptoms and integrated management of any two of the following cotton diseases :
   i. Southern and Northern blight
   ii. Banded leaf and sheath blight
   iii Northern blight
5. Illustrate the downy mildew of maize in the following headings :
   (i) Pathogen
   (ii) Symptoms
   (iii) Disease cycle
   (iv)Disease management
6. Describe in detail the most distinguishing symptoms, causal organism, disease cycle and management of northern blight of maize.

b. Short Answer Questions

1. How will you differentiate Southern and Northern blight of maize?
2. Why the diseases is called Southern and Northern blight of maize?
3. Which fungal pathogen causes downy mildew of maize? Give its systematic position.
4. Write a short note on Race-T of *Helminthosporium maydis.*
5. Describe the mode of parasitism of *Peronosclerosporasorghi.*

c. Very short answer Questions

1. Suggest some cultural practices for management of downy mildew disease of maize.
2. List three important diseases of maize and their causal organisms.
3. Distinguish between *Helminthosporium maydis* and *Helminthosporium turcicum*.
4. Which type of disease is can be effectively managed by mixed cropping.
5. Discuss in brief about chemical mangment of Southern and Northern blight of maize

Fig. 1: Banded leaf and sheth blight

Fig. 2: Southern leaf blight

Fig. 3: Northern leaf blight

Fig. 4: Downy mildew

Plate 3: Photograph showing symptoms of major diseases of maize

4

Sorghum Crop Diseases & Management

Major Diseases	Causal Organism
Loose smut	*Sphacelotheca cruenta*
Long smut	*Tolyposporium ehrenbergii*
Head smut	*Sphacelotheca reiliana*
Grain smut	*Sphacelotheca sorghi*
Grain mold	Fusarium, Curvularia, Alternaria etc
Anthracnose	*Colletotrichum graminicolum*
Ergot or Sugary	*Sphacelia sorghi*
Downy mildew	*Peronosclerospora sorghi*
Leaf blight	*Exerohilum turcicum*
Rectangular leaf spot	*Cercospora sorghi*
Rust	*Puccinia purpurea*
Phanerogamic parasite	*Striga asiatica* and *Striga densiflora*

Loose Smut

Sphacelotheca cruenta

Diagnostic Symptoms

- Affected plants can be detected before the ears come out.
- Shorter than the healthy plants with thinner stalks and marked tillering.
- Ears come out much earlier than the healthy.
- Glumes are hypertrophied and the ear head gives a loose appearance than healthy.
- Sorus is covered by a thin membrane which ruptures very early, exposing the spores even as the head emerges from the sheath(Fig-1).

Etiology

Scientific Classification

Kingdom : Fungi

Division : Eumycota

Sub- division : Basidiomycotina

Class : Teliomycetes

Order : Ustilaginales

Family : Ustilaginaceae

Genus : *Sphacelotheca*

Species : *cruenta*

- Mycelium is multicellular.
- Spore mass is dark brown , surrounding a wall developed , unbranched columella of host tissue.
- Spores are globose to subglobose, light yellowish brown, wall 1 micron thick, finely echinulate, 6-12 micron in diameter.
- Spherical or ellipsoidal, tinted, smooth teliospores are 4.8 x 10.8 µm, rarely to 12 µm in diameter.
- The spores germinate readily in water or nutrient solutions producing four-celled basidium with laterally and apically growing sporidia.
- Sporidia are spindle-shaped or oblong, 2.8 x 12.7 µm in size.
- Budding of sporidia is common.

Disease Cycle

- Pathogen is mainly externally seed borne.However, it is stated to be soil borne to some extent in dry soils.
- Spores remain viable for four years in the laboratory, but as they germinate readily in water soil transmission in not very important.
- Spread within the crop by floral infection by air spores has been observed.

Epidemiology

- Soil temperature- 18-23 ^{0}C.
- Soil humidity- 15-20%.

Management

- Deep summer ploughing.
- Use disease free seeds.

- Careful removal of infected earheads.
- Avoid successive cultivation of a single variety/hybrid in a specific area.

Long Smut

Tolyposporium ehrenbergii

Diagnostic Symptoms

- Disease is generally restricted to a relatively a small proportion of the florets which are scattered on a head.
- Sori are long, cylindrical, elongated, slightly curved with a relatively thick creamy-brown covering membrane (peridium).
- Peridium splits at the apex to release black mass of spores.

Etiology

Scientific Classification

Kingdom : Fungi

Division : Eumycota

Sub- division : Basidiomycotina

Class : Teliomycetes

Order : Ustilaginales

Family : Ustilaginaceae

Genus : *Tolyposporium*

Species : *ehrenbergii*

- Spore mass is granular, black, intermixed with shreds of host tissue.
- Spore balls are many spored, permanent, rather irregular in size and shape, globose to elongated, dark brown in patches, opaque, 45micron or less, with a maximum diameter of more than 200 micron.
- Spores at the surface of the ball are dark brown with the free surface having papillae.The inner spores are pale in color, smooth, globose to subglobose or angular and 10-15 micron in diameter(Fig-2).

Disease Cycle

- Floral infection occurs by means of sporidia which enter the floral parts, producing sori about 12 to 15 days later.
- Spores in the soil may germinate to produce clusters of sporidia which become air borne and cause primary infection.
- Secondary infection may also take place through the spores released from the smutted ears in the field.

Management

- Follow sanitary practices to reduce the infection potential are advisable.
- Follow crop rotation.

Head Smut

Sphacelotheca reiliana

Diagnostic Symptoms

- Disease is observed at flowering stage.
- Entire head is replaced by large **sori**.
- Sorus is covered by a whitish grey membrane of fungal tissue, which ruptures, before the head emerges from the boot leaf to expose a mass of brown smut spores.
- Spores are embedded in long, thin, dark colored filaments which are the vascular bundles of the infected head(Fig-3).

Etiology

Scientific Classification

Kingdom : Fungi

Division : Eumycota

Sub- division : Basidiomycotina

Class : Teliomycetes

Order : Ustilaginales

Family : Ustilaginaceae

Genus : *Sphacelotheca/Sporisorium*

Species : *reilianum*

- Pathogen forms both teliospores (hibernating stage of the pathogen) and sporidia in its life cycle.
- Teliospores: Spherical, roundish, angular, ellipsoidal, always setaceous, tinted, 12.1 to 13.7 μm in diameter teliospores are gathered in easily disintegrating glomerules. They sprout into three or five-celled basidia with laterally and apically growing sporidia.
- Sporidia are spindle-shaped or oblong, 3.2 to 10.3 μm in size, with multicellular mycelium.

Disease Cycle

- Infection takes place at the time of seed germination and seedling emergence through the smut spores adhered to the seed.
- Seed spores adhered to seed retain viability for two years.
- Smut spores germinate on seed coat and gives rise sporidia and infect embryonic growing point.
- No infection takes place once the primary shoot comes out from soil.
- Pathogen follow the growth of the host, mycelium rapidly branches in the crown bud and floral structure.
- Spores are produced by the cells of hyphae, which replace the kernel.

Epidemiology

- Optimum temperature - 14 to 36°C.
- Soil temperature- 18-23°C.
- Soil humidity- 15-20%.

Management

- Use disease free seeds.
- Shallow sowing makes the seedlings escape the infection.
- Seed with Captan or Thiram at 4 g/kg.
- Follow crop rotation.
- Collect the smutted ear heads in cloth bags and bury in soil.

Kernel smut/ Covered smut /Grain smut / Short smut

Sphacelotheca sorghi

Diagnostic Symptoms

- Disease appears only at the time of grain formation in the ear.
- Individual grains are replaced by smut sori. They are generally larger than the normal grains.
- Sori are oval or cyclindrical and are covered with a tough creamy skin (peridium) which often persists unbroken up to thrashing(Fig-4).

Etiology

Scientific Classification

Kingdom : Fungi

Division : Eumycota

Sub- division : Basidiomycotina

Class : Teliomycetes

Order : Ustilaginales

Family : Ustilaginaceae

Genus : *Sphacelotheca*

Species : *sorghi*

- Pathogen in its development cycle forms teliospores (wintering stage of the fungus) and sporidia.
- Teliospores are spherical or roundish, tinted, smooth, 4.8 to to 8.5 m in diameter.
- At sprouting they form three- or four-cellular basidium with laterally and apically growing sporidia.
- Sporidia are oblong, with slightly rounded ends, or spindle-shaped, 2.9 to 18.5 μm.

Disease Cycle

- Smut spores are blown by wind gets adhere on seed coat.
- If the spore sac wall is broken, the spores are adhered with the seeds during threshing of seeds.

- Seed lodged spores remain dormant until next season.
- Seedling infection takes place at the time of seed germination and emergence either by germinated smut spores lodged with seeds or present in the soil with other seed.
- Smut spores produce sporidia or infection hyphae and cause infection. They reach near the budding point and remain viable.
- Pathogen follows the growth of the host and mycelium branches in the crown bud.
- Spores are produced by the fungal mycelium by replacing the kernel.

Epidemiology

- Optimum temperature - 10 to 32°C.
- Soil temperature- 18-23°C.
- Soil humidity- 15-20%.

Management

- Deep summer ploughing.
- Use disease free seeds.
- Careful removal of infected earheads.
- Avoid successive cultivation of a single variety/hybrid in a specific area.

A comparison of the chief characters of the four smut diseases is presented in the table

Table Comparison of the four smuts of Sorghum

Characters	Head smut	Long smut	Loose smut	Grain smut
Pathogen	*Sphacelotheca reiliana*	*Tolyposporium ehrenbergii*	*Sphacelotheca cruenta*	*Sphacelotheca sorghi*
Host	Heading premature, Not stunted	Heading normal, Not stunted	Heading normal, Not stunted	Heading premature, Not stunted
Ear Infection	Entire head smutted	Very few grains are infected	Most or all grains smutted	Most or all grains smutted
Site	Inflorescence	Ovary	Ovary	Ovary
Sori	Very big	Long	Small	Small
Membrane	Ruptures easily	Relatively thick membrane	Rupture easily	Tough and persists
Columella	Absent however, a network of vascular tissue present	Absent	Present	Present
Spores	Loosely bound into balls, spherical or angular, dull brown, minutely papillate 10-16 micron in diameter	Always in balls, globose or angular, brownish green, warty spore wall, 12-16 micron in diameter.	Single spherical or elliptical, dark brown, spore walls, pitted, 5-10 micron in diameter.	Single round to oval, olive-brown, smooth walled, 5-9 micron in diameter.
Viability of spores	Up to 2 years	About 2 spores	About 4 years	Over 10 years
Survival	Soil and seed borne	Air borne	Externally seed borne	Externally seed borne

Grain Mould

- A serious issue that jeopardizes the nutritional content, safety, and quality of sorghum grain when used as pasture for animals and human use.
- The complex of diseases known as sorghum grain mold, which poses a serious threat to human health in addition to endangering agricultural profitability and productivity by producing toxic mycotoxins..

Diagnostic Symptoms

- Also called Head blight /Head mould.
- Pigmentation of the lemma, palea, glumes, and lodicules is one of the initial signs that are noticeable.
- The symptoms might vary greatly depending on the type of fungus implicated, the stage of grain maturity, and the extent of infection.
- While partially infected grain may appear normal and discolored, severely infected grain is completely coated in mold.
- Some grain that appears normal may not exhibit any visible signs, but following surface sterilization, it develops mold fungus on a blotter.
- Fungal growth starts at the grain's hilar end and spreads to the pericarp surface after that.
- When sorghum grains are severely infected, the resulting multicolored grains are caused by different colored fungal mycelium and sporulating structures, which vary depending on the type of fungal pathogen that is invading the grains.
- Fungal infection causes more noticeable grain discoloration on white-grain sorghums than on brown or red-grain varieties.
- Small, shriveled grains on the panicle, poor seed set, and blasted florets are the effects of early infection at anthesis.
- Grains with severe infection crumble even with mild pressure(Fig-5).
- Head blight disease, which causes extensive damage of panicle, including peduncle and rachis branches and spikelets, to become infected with Fusarium spp. during the early stages of flowering, is known as blighted panicle.
- Depending on the type of fungus involved, the color and texture of grains colonized by mold fungi can differ.

- *Curvularia lunata* is the fungus that colonizes the fastest emerges on the grain surface as a glossy, velvety black, fluffy growth.
- The majority of Fusarium species develop pinkish white mycelium, which initially has a powdery appearance before changing into pinkish fluffy (*F. pallidoroseum*) and fluffy white (*F. verticillioides*).
- *Phoma sorghina* yields a thick, dirty, black crust with a rough surface on the pericarp, as well as tiny, round, black pycnidia that resemble pins embedded in grain .
- The appearance of *Alternaria alternata* is drab, with grayish black mycelium that is frequently sparse and striped .
- On the grain surface, *Bipolaris australiensis* appears as dark black mycelium.
- In contrast to *Colletotrichum graminicola*, which develops small, black, expanded acervuli studded with clusters of setae forming concentric rings on the grain, *Cladosporium oxysporum* grows as grayish mycelium with a powdery appearance on the grain .
- Grain in humid conditions takes on a "moldy appearance" because to post-mature colonization.Such surface molding does not compromise grain integrity, but it lowers the grain's market value because of its moldy appearance.
- Rainfall during crop maturity increases the likelihood that grains will show signs of mold and perhaps sprout in the panicle.
- Pathogenic and saprophytic phases at various stages of grain formation can occasionally be challenging to navigate.
- These fungi have extremely variable growth and sporulation patterns that are sensitive to variations in the weather.

Etiology

- The main fungi responsible for the grain mold complex in India include *Alternaria alternata, Phoma sorghina, Curvularia lunata, F. verticillioides, and F. thapsinum.*
- The most pathogenic of them are *F. verticillioides, C. lunata*, and *A. alternate.*
- Throughout the cropping season, these fungi occur more or less frequently depending on the site and surrounding circumstances.

- The majority of other fungi were saprophytes and aid in the weathering of post-maturity grains. These fungi included *Aspergillus niger, A. fumigatus, Cladosporium* spp., Bipolaris spp., Epicoccum spp., *Fusarium semitectum*, and Olpitrichum spp.

Disease Cycle

- The fungi survive as parasites as well as saprophytes in the infected plant debris.
- The fungi mainly spread through air-borne conidia.

Epidemiology (Weather variables and mold development)

- Temperature and relative humidity have a significant impact on the growth of mold and the infection of grain mold fungus.
- During the stages of grain development and flowering, warm, humid circumstances encourage the growth of mildew and infections, whilst dry conditions inhibit them.
- The frequency of fungal infection rose as wetness duration increased up to 72 hours.
- The frequency of infection varied among the fungus for different growth phases at which the panicles were injected, suggesting that different fungi may have distinct windows for maximum infection during the stages of grain development.
- The effects of panicle wetness periods on *P. sorghina, F. verticillioides*, and *C. lunata* infection were highly significant.

Management

- Adjust the sowing time.
- Spray any fungicides like Mancozeb 1 kg/ha or Captan 1 kg + Aureofungin-sol 100 g/ha. in case of intermittent rainfall during earhead emergence, a week later and during milky stage.

Anthracnose and Red Rot

Colletotrichum graminicolum

Diagnostic Symptoms

- Initially small red coloured spots appear on both surfaces of the leaf.

- Centre of the spot is white in colour encircled by red, purple or brown margin.
- Numerous small black dots like acervuli are seen on the white surface of the lesions.
- Red rot can be characterized externally by the development of circular cankers, particularly in the inflorescence.
- Infected stem when split open shows discoloration, which may be continuous over a large area or more generally discontinuous giving the stem a marbeled appearance.
- Pathogen causes both leaf spot (anthracnose) and stalk rot (red rot) (Fig-6).

Pathogen

Scientific Classification

Kingdom : Fungi

Division : Eumycota

Sub- division : Deuteromycotina

Class : Coelomycetes

Order : Melanconiales

Family : Melanconiaceae

Genus : *Gloeosporium*

Species : *ampelophagum*

- Mycelium of the fungus is localized in the spot.
- Acervuli with setae arise through epidermis.
- Conidia are hyaline, single celled, vacuolate and falcate in shape.

Disease Cycle

- Pathogen survives in infected plant debris.
- Disease spread by means of seed-borne and air-borne conidia.

Epidemiology

- Temperature -28-30°C.
- High humidity.
- Continuous rain.

Management

- Seed treatment with Captan or Thiram at 4 g/kg.
- Foliar spray with Mancozeb @ 0.2%, if required repeat after fifteen days.

Model Question Paper

A Objective Type Questions

a. Choose the correct answer from the following

(1) Which of the following is an externally seed borne disease.

a. Head smutof sorghum b. Long smutof sorghum

c. Loose Smut of sorghum d. None of the above

(2) Kernel smut of Sorghum is caused due to

a. *Sphacelotheca reiliana*

b. *Tolyposporium ehrenbergii*

c. *Sphacelotheca sorghi*

d. *Sphacelotheca cruenta*

(3) Which of the following smut disease is air borne in sorghum ?

a. Head smut of sorghum b. Long smut of sorghum

c. Loose smut of sorghum d. None of the above

(4) Head smut of sorghum was reported for the first time in.

a. Russia b. India

c. USA d. Pakistan

(5) *Sphacelotheca reiliana* is the causal organism of

a. Head smut of sorghum b. Long smut of sorghum

c. Loose smut of sorghum d. Grain smut of sorghum

(6) The survival of *Sphacelotheca reiliana* is:

a. Soil and seed borne b. Air borne

c. Externally seed borne d. Internally seed borne

(7) The site of infection of head smut pathogen*Sphacelotheca reiliana* is:

a. Inflorescence b. Ovary

c. Leaf d. Root

(8) Very few grains are infected in which smut disease of sorghum

a. Head smutof sorghum b. Long smut of sorghum

c. Loose smut of sorghum d. Grain smut of sorghum

(9) The ideal soil temperature for head smut of sorghum is:

a. 18–23°C b. 14–25°C

c. 18–30°C d. 20–25°C

(10). Sugary disease is related to which crop

a. Sorghum b. Rice

c. Wheat d. Chickpea

Answer

S. No	Answer	S. No	Answer
1	c.Loose smut of sorghum	6	a. Soil and seed borne
2	c. *Sphacelotheca sorghi*	7	a. Inflorescence
3	b. Long smut of sorghum	8	b.Long smut of sorghum
4	a.Russia	9	a.18–23°C
5	a.Head smut of sorghum	10	a.Sorghum

b Fill in the blanks with suitable word (s)

1. Loose smut of sorghum is caused by ....
2. *Sphacelia sorghi* causes .... disease.
3. Of the four smut diseses in sorghum .... disease is air borne.
4. *Tolyposporium ehrenbergii* is the causal organism of ....
5. Smuts are obligate parasite, except ----------------------, which grows well in the soil.
6. Smut is a German word means---------------.
7. Smut diseases caused by *Sphacelotheca sorghi* is called------------------.
8. Acervuli with setae arise through epidermis in -----------------------disease of sorghum.

9. Loose smut of wheat and loose smut of ---------------------are same except in host specificity.
10. The pathogen *Sphacelia sorghi* in its development cycle forms ---------- and sporidia.

S. No	Answer	S. No	Answer
1	*Sphacelotheca cruenta*	6	Dirty
2	Grain smut	7	Grain smut
3	Long smut	8	Anthrcnose
4	Long smut	9	Barley
5	Ustilago hordei	10	Teliospore

(c) State whether the following statements are *True* or *False*

1. Ergot poses a significant obstacle to the production of hybrid seeds in sorghum.
2. Kernel smut disease appears only at the time of grain formation in the ear of sorghum.
3. Loose smut infected plants can be detected before the ears come out in sorghum.
4. In sugary disease, the infection is identified by the exudation of pink droplets from the diseased ovary.
5. The conidia of *Sphacelia sorghi* have a falcate, vacuolate, single-celled, hyaline form.
6. The budding of sporidia is common in *Sphacelotheca cruenta.*
7. The pathogen *Sphacelotheca cruenta* is mainly externally seed borne.
8. Long smutdisease is generally restricted to a relatively a small proportion of the florets which are scattered on a head of sorghum .
9. The entire head is replaced by large sori in head smut of sorghum.
10. Head smut disease is observed at flowering stage in sorghum.

Answer

S. No	Answer	S. No	Answer
1	True	6	True
2	True	7	True
3	True	8	True
4	True	9	True
5	True	10	True

B. Descriptive Questions

a. Long answer questions

1. How do the four smut of sorghum differ? Describe in details.
2. Describe the loose smut of sorghum under heads, diagnostic symptoms, causal organism, disease cycle and disease management.
3. Which are the important smut diseases of sorghum ? Describe any one under heads, diagnostic symptoms, causal organism, disease cycle and disease management.
4. Distinguish between
   a) Localized and Systemic smut infection
   b) Monocyclic and polycyclic smut diseases
   c) Long smut and Short smut
5. Write short notes on
   a. Economi c importance of smut of sorghum in India
   b. Management of smut disease of sorghum
   c. Disese cycle of loose smut of sorghum
6. Briefly describe the diagnostic symptoms and management of the following plant disease
   a. Ergot or sugary of sorghum
   b. Loose smut of sorghum
   c. Kernel smut of sorghum.
7. Describe in detail the symptoms, causal organism, disease-cycle and management of head smut of sorghum.
8. Write the causal organism and major symptoms of any three of the following plant disease:
   (i) Ergot disease of sorghum
   (ii) Loose smut of sorghum
   (iii) Covered smut of sorghum
9. Describe the symptoms and disease management of any two plant diseases caused by the following pathogens:

(i) *Sphacelotheca cruenta*

(ii) *Tolyposporium ehrenbergii*

(iii) *Sphacelotheca reiliana*

10. Describe in detail the symptoms, causal organism, disease-cycle and management of long smut of sorghum

b. Short answer questions

1. Differentiate between loose smut and head smut of sorghum.
2. How would you identify the following disease in the field

   (a) Loose smut of sorghum

   (b) Sugry disease of sorghum
3. Write the epidemiological condition for development of covered smut of sorghum .
4. Discuss in brief about management of loose smut of sorghum.
5. Write down the integrated management schedule of Long smut of sorghum.

c. Very short answer Questions

1. Name the pathogen or causal organism of following plant disease

   (i) Ergot disease of sorghum

   (ii) Loose smut of sorghum

   (iii) Covered smut of sorghum

   (iv)Long smut of sorghum
2. Explain cultural management of smut of sorghum.
3. Write down the integrated management schedule of long smut of sorghum
4. Differentiate between smut and rust.
5. Name the three smut diseases of sorghum with their causal organism.

Fig. 1: Loose smut

Fig. 2: Long smut

Fig. 3: Head smut

Fig. 4: Grain smut

Fig. 5: Grain mould

Fig. 6: Anthracnose

Plate 4: Photograph showing symptoms of major diseases of Sorghum

5

Bajra Crop Diseases & Management

Major Diseases	Causal Organism
Downy mildew	*Sclerospora graminicola*
Ergot or Sugary	*Claviceps fusiformis*
Smut or Grain smut	*Tolyposporium penicillariae*
Rust	*Puccinia pennisetti*

1. Downy Mildew (Green Ear)

Sclerospora graminicola

Diagnostic Symptoms

- Characteristic symptoms of the disease are pale, chlorotic, broad streaks extending from base to tip of leaves.
- At the advancement of disease, the leaf streaks turn brown and the leaves become shredded longitudinally.
- In severe infection, the downy fungal growth can be seen on the upper as well as lower surface of the leaves.
- Infected plants fail to form ear but if formed, they are malformed to green leafy structures.
- Complete ear can be transformed into leafy structure.
- Pathogen transformed all floral parts such as glumes, palea, stamens and pistils into green linear leafy structures of variable lengths and no grains are produced.
- Partial infection of ear head is observed in which some grains are produced and rest of the ear head is converted in to a green ear(Fig-1).

Etiology

Scientific Classification

Kingdom : Fungi

Division : Eumycota

Class : Oomycetes

Order : Peronosporales

Family : Peronosporaceae

Genus : *Sclerospora*

Species : *graminicola*

- Mycelium is intercellular, non septae and systemic.
- Short, stout, hyaline sporangiophores arise through stomata and branch irregularly with stalks bearing sporangia.
- Sporangia are hyaline, thin walled, elliptical and bear prominent papilla.
- Oospores are round in shape, surrounded by a smooth, thick and yellowish brown wall.

Disease Cycle

- Oospores remain viable in soil for 5 years.
- Primary seedling infection takes place through the germinating oospores already present in the soil resulting in systemic infection.
- Oospore are generally transmitted through the infected seeds and plant residues in soil.
- Secondary spread is through wind borne sporangia produced during rainy season.
- Dormant mycelium of the fungus is present in embryo of infected seeds.

Epidemiology

- Atmospheric temperature -15-25 °C.
- Relative humidity above 85 per cent.
- Light drizzling accompanied by cool weather.

Management

- Deep ploughing to reduce population of pathogen in soil.
- Use disease free seeds and avoid seeds from diseased field.
- Apply 25 kg N/ha 25 days after sowing.

- Grow resistant varieties ICTP-8203,AIMP-92901(Samrudhi), PPC-6 (Sampada) and hybrids like Shradha, Saburi and Pratibha.
- Seed treatment with Metalaxyl at 6g/kg or with Trichoderma or Pseudomonas @10g/kg of seed.
- Foliar spray with Mancozeb (0.2%) or Metalaxyl + Mancozeb (0.2%) on 15th day after sowing in the field.
- Roguing out infected plants.
- Follow intercrop with pigeonpea in 4:2 ratio.

2. Ergot or Sugary

Claviceps microcephala

Diagnostic Symptoms

- Cream to pink mucilaginous droplets of "honeydew" ooze out of infected florets and form sclerotia.
- Within 10 to 15 days, the droplets dry and harden, and dark brown to black sclerotia develop in place of seeds on the panicle.
- Sclerotia are larger than seed and irregularly shaped, and generally get mixed with the grain during threshing(Fig-2).

Etiology

Scientific Classification

Kingdom : Fungi

Division : Eumycota

Sub- division : Ascomycotina

Class : Plectomycetes

Order : Clavicipitales

Family : Clavicipitaceae

Genus : *Claviceps*

Species : *microcephala*

- Pathogen produces septate mycelium which produces conidiophores and is closely arranged.

- Conidia are hyaline and one celled.
- Sclerotia are small (3-8mm x 0.3-15mm) and dark grey.

Disease Cycle

- Primary infection takes place by germinating sclerotia present in the soil or through the sclerotia admixture with seeds.
- Sclerotia are viable in soil for 6-8 months.
- Sclerotia take about 30-45 days to germinate and produce air borne spores which spread primary infection to pearl millet crop.
- Secondary spread of the disease is through conidia produced in large numbers in honey dew and disseminated by insects or rain.
- Role of collateral hosts like *Cenchrus ciliaris* and *C. setigerus* in perpetuation of fungus is significant.

Epidemiology

- Relative humidity - > 80%.
- Temperature-20 to 30 ^{0}C.
- Intermittent rains and dry spell trigger sclerotial germination and ascospore discharge.

Management

- Use clean ergot free seeds.
- Immerse the seeds in 10 per cent salt solution and remove the floating sclerotia.
- Adjust the sowing date so that the crop does not flower during September when high rainfall and high relative humidity favour the disease spread.
- Deep ploughing reduces the viability of sclerotia.
- Heads which emerges quite early could be removed if found having honey dew stage of infection.
- Remove collateral and alternate hosts.
- Foliar spray with Carbendazim (0.1%) or Mancozeb (0.25%) when 5-10 per cent flowers have opened and repeat at 50 per cent flowering stage.

Model Question Paper

A. Objective Questions

a. Choose the correct answer from the following

(1) Down mildew of bajra also known as .

a. Green ear b. Brown spot

c. False smut d. Tundu

(2) Down mildew of bajra is caused due to :

a. *Ustilaginoidea virens* b. *Sclerospora graminicola*

c. *Tilletia indica* d. *Clavibacter tritici*

(3) Complete ear can be transformed into leafy structure in which disease of bajra

a. Downy mildew b. Grain smut

c. Ergot d. None of all

(4). Sporangia of *Sclerospora graminicola* germinate producing bi-flagellate, reniform

a. zoospores b. Oospore

c. Aplanospores d. None

(5) The downy mildew disease of bajra is mainly

a. Soil borne b. Seed borne

c. Air borne d. Water borne

(6) The optimum atmospheric temperature for development of downy mildew of bajra is

a. 15-25°C b. 35-40°C

c. 10-15°C c. 35-⁻45°C

(7). The honeydew stage is diagnostic symptom of

a. Blast of rice b. Brown spot of rice

c. Rust of bajra d. Ergot of bajra

(8) The pathogen *Claviceps microcephala* belongs to class

a. Plectomycetes b. Phycomycetes

c. Ascomycetes d. Bsidiomycetes

Answer

Sl.No	Answer	S. No	Answer
1	a. Green ear	6	a.15-25 °C
2	b.*Sclerospora graminicola*	7	d.Ergot of bajra
3	a.Downy mildew	8	a. Plectomycetes
4	a.zoospores		
5	a.Soil borne		

b. Fill in the blanks with suitable word (s)

1. ----------- of *Sclerospora graminicola* remain viable in soil for 5 years.
2. Primary seedling infection of *Sclerospora graminicola* takes place through the germinating ------------------.
3. *Sclerospora graminicola* belongs to the class ----------------.
4. The cream to pink mucilaginous droplets of "honeydew" ooze out of ergot infected florets of bajra and form ------------------.
5. The sclerotia *Sclerospora graminicola* are viable in soil for -------- months.

Answer

Sl.No	Answer	S. No	Answer
1	Oospores	4	Sclerotia
2	Oospores	5	6-8
3	Oomycetes		

c. State whether the following statements are *True* or *False*

1. Partial infection of ear head is observed bajra infected with downy mildew disese in which some grains are produced and rest of the ear head is converted in to a green ear.
2. The sporangia of *Sclerospora graminicola* are hyaline, thin walled, elliptical and bear prominent papilla.
3. The oosporic stage of the *Sclerospora graminicola* is most common and predominant stage in India.

4. The dormant mycelium of the *Sclerospora graminicola* causing downy mildew of bajra is present in embryo of infected seeds.
5. The pathogen *Claviceps microcephala* attacks the ovary and grows profusely producing masses of hyphae .
6. Deep ploughing reduces the viability of sclerotia produced by *Claviceps microcephala.*
7. Sclerotia of *Claviceps microcephala* take about 30-45 days to germinate and produce air borne spores which spread primary infection to pearl millet crop.
8. Secondary spread of the ergot of bajra is through conidia produced in large numbers in honey dew and disseminated by insects or rain.

Answer

S. No	Answer	S. No	Answer
1	True	6	True
2	True	7	True
3	True	8	True
4	True		
5	True		

B. Descriptive Questions

a. Long answer questions

1. Briefly describe the diagnostic symptoms, causal organism, disease cycle and disease management of the following plant disease
   a. Downy mildew
   b. Ergot or Sugary
2. Describe in detail the symptoms, causal organism, disease-cycle and management of downy mildew of bajra.
3. Write the causal organism and major symptoms of any three of the following bajra crop disease:
   a. Downy mildew
   b. Ergot or Sugary
   c. Rust.

4. Describe the symptoms and disease management of any two plant diseases caused by the following pathogens:

   (i) *Sclerospora graminicola*

   (ii) *Claviceps fusiformis*

   (iii) *Puccinia substriata*

5. Describe the ergot disease bajra under heads, diagnostic symptoms, causal organism, disease cycle and disease management.

6. Write down the major fungal diseses of bajra? Describe any one in detail under heads, diagnostic symptoms, causal organism and integrated management.

b.Short answer questions

1. Differentiate between rust and smut.

2. How would you identify the disease in the field

   (a) Downy mildew of bajra

   (b) Ergot of Bajra

3. Write the epidemiological condition of downy mildew of bajra.

4. Write short notes on integrated management of downy mildew of bajra.

5. Write short notes on etiology of ergot of bajra.

Fig. 1: Downy mildew

Fig. 2: Ergot or Sugary

Plate-5: Photograph showing symptoms of major diseases of bajra

6

Finger Millet Crop Diseases & Management

Major Diseases	Causal Organism
Blast	*Pyricularia grisea*
Leaf spot	*Cercospora eleusine*

1. Blast

Pyricularia grisea

Diagnostic Symptoms

- Crop is susceptible to the disease during all stages of its growth, from seedling to grain formation stage.
- In young seedlings, lesions are generally of spindle shaped, however lesions of different sizes may also be observed.
- In the beginning the spots have yellowish margin and grayish centre. Later, the centres become whitish grey and disintegrate.
- Lesions are isolated in the beginning and afterwards they may soon coalesce. The distal portion of the leaves beyond the lesions may hang and drop off.
- In older plants, the disease appears on leaf blades with typical spindle shaped spots.
- Under congenial conditions, such spots enlarge, coalesce and form bigger lesions.
- Lesions are like those on the seedlings and are about 0.3-0.5 cm in breadth and 1-2 cm in length. Apices of the infected leaves beyond the lesions hang down and sometimes break.
- Stem infection causes blackening of the nodal region, penetrating into the tissues.

- Maximum damage is caused by neck infection. The neck region turns black and shrinks.
- During severe infection the ears hang down from the stalk and sometimes may also break away(Fig-1).

Etiology

Scientific Classification

Kingdom : Fungi

Division : Eumycota

Sub- division : Ascomycotina

Class : Sordariomycetes

Order : Magnaporthales

Family : Magnaporthaceae

Genus : *Magnapotthe*

Species : *grisea*

- Ragi yeast lactose agar was the best medium for growth and sporulation.
- Colour of aerial mycelium varied from creamy white to greyish black with circular colony and crenate edge.
- Growth pattern - Fluffy and vulvetty.
- Mycelium is intra and intercellular.
- Numerous conidiophores and conidia are produced in the central portion of the spindle shaped spot under humid condition.As a result the spot assumes smoky appearance.
- Conidiophore emerges through stomata or through epidermal cells are simple septate and dark colored.
- Conidia are borne at the tips of the conidiophores.
- Conidia are pyriform , three celled ,hyaline , two spetate.
- Conidia germinate with germtube which infect the leaf wither through epidermal cells or stroma.

Disease Cycle

- Pathogen harbours in glumes, straw as well as on some graminaceous weeds.
- Seeds have also been found to carry the pathogen in the pericarp and endosperm.

Epidemiology

- Temperature -25 to 30°C.
- Relative humidity-more than 90 per cent.
- Cloudy days with intermittent rainfall.
- Continuous rains.
- Increased doses of nitrogen.

Management

- Seed treatment with Tricyclazole (8g/kg seed).
- Two sprays of Ediphenphos or Kitazin or Propiconazole (0.1%) or Carbendazim or Tricyclazole (0.05%), first at the time of ear emergence and second after 10 days or an initial spray of 0.05% Carbendazim or Tricyclazole followed by a spray of Mancozeb (0.2%) 10 days later are effective.
- Two sprays of Mancozeb+ Carbendazim (0.2%) first at 50 per cent flowering and again 10 days later.
- Use of resistant varieties/genotypes like; VL 149, GPU 26, GPU 28, GPU 45, CO 13, GE 632, GE 637, GE 669, GE 674, GE 676, GE 705, GE 718, GE 728, 2400, 4913, 4914, 4915, 4929, 4966, 5102, 5126 and 5148, IE 287, IE 976 and IC 43335.
- Seed treatment with *Trichoderma harzianum* (TH) and one spray of *Pseudomonas fluorescens* at the time of flowering and second spray 10 days later can also control all three blast diseases.

2. Cercospora Leaf Spot

Cercospora eleusine

Diagnostic Symptoms

- Disease occurs most severely in the month of June in the early sown crop.

- Infection generally starts from the older leaves and then spreads to the young leaves.
- Thus progressive infection from older to younger leaves can be observed in the standing crop.
- Initial symptoms appear as reddish brown speaks with yellow halo and are easily confused with those of Helminthosporiose.
- Later, several such specks coalesce to form large lesions with any yellow hallow.
- In some cases the lesions enlarge to assume eye shaped spot measuring up to 15x3mm. Such leaves give burnt appearance.
- At the time of crop maturity severely infected leaves turn completely necrotic, shrivel and dry.
- At this stage the plants look completely blighted. Such symptoms can also be seen on the stem, leaf sheath and fingers(Fig-2).

Scientific Classification

- Kingdom : Fungi
- Division : Eumycota
- Sub- division : Deuteromycotina
- Class : Hyphomycetes
- Order : Hyphomycetales (Moniliales)
- Family : Dematiaceae
- Genus : *Cercospora*
- Species : *eleusine*
- Mycelium is intra and intercellular, septate, light brown in color.
- Conidia are erect or curved septate and dark brown.They are borne at the tip of conidiophores.
- Conidiophore are thick walled, cylindrical, obclavate, straight or curved, and light brown in color and 3-10 septate.
- The spores germinate either through stomata or epidermal cells.
- Eleven spores are formed in one conidiophore.

Disease Cycle

- Little is known about the disease cycle, inoculum sources, or how the pathogen overwinters.
- The fungus may survive between millet crops in and on infected crop residue as spores (conidia) and fungal mycelial masses (stromata), weeds, and perhaps on alternate hosts.
- The pathogen is likely spread by splashing rain and irrigation water and, to a lesser degree, by wind.

Epidemiology

- Disease emergence occurs when high temperatures coincide with periods of high humidity.
- High rainafall.

Management

- Field sanitation.
- Avoid planting millet varieties that are highly susceptible to Cercospora fungi; control weeds in field.
- Rotate crops and practice good sanitation.
- Foliar spray of Carbendazim @ 0.1% or Mancozeb @ 0.25%.

Model Question Paper

A. Objective Type Questions

a. Choose the correct answer from the following

(1) The best medium for growth and sporulation of*Magnaporthe grisea*

a. Ragi yeast lactose agar medium

b. Yeast extract manitol agar medium

c. Potato dextrose agar medium

d. None

(2) Blast disease of finger millet caused due to :

a. *Pyricularia grisea* b. *Sclerospora graminicola*

c. *Tilletia indica* d. *Clavibacter tritici*

(3) The disease appears on leaf blades with typical spindle shaped spots on finger millet crop

a. Blast b. Grain smut

c. Leaf spot d. None of all

(4) The pathogen *Magnaporthe grisea* belongs to order

a. Magnaporthales b. Peronosporles

c. Chytridiales d. None

(5) The blast disease of fingermillet is mainly

a. Soil borne b. Seed borne

c. Air borne d. Water borne

(6) The optimum atmospheric temperature for development of blast of fingermillet is

a. 25-30 °C b. 35-40^{0}C

c. 10-15^{0}C c. 35-$^-$45^{0}C

(7) The pathogen *Cercospora eleusine* belongs to class

a. Hyphomycetes b. Phycomycetes

c. Ascomycetes d. Basidiomycetes

Answer

Sl.No	Answer	Sl.No	Answer
1	a. Ragi yeast lactose agar medium	5	b.Seed borne
2	a.*Pyricularia grisea*	6	a.25-30 °C
3	a. Blast	7	a. Hyphomycetes
4	a.Magnaporthales	8	-

b. State whether the following statements are *True* or *False*

1. *Magnaporthe grisea* belongs to order Magnaporthales.
2. The colour of aerial mycelium *of Magnaporthe grisea* from creamy white to greyish black with circular colony and crenate edge.
3. Seeds have also been found to carry the pathogen *Magnaporthe grisea* n the pericarp and endosperm.
4. Seed treatment with Tricyclazole @8g/kg seed control blast of fingermillet.

5. The symptom of Cercospora leaf spot generally starts from the older leaves and then spreads to the young leaves.
6. The initial symptoms of Cercospora leaf spot appear as reddish brown speaks with yellow halo and are easily confused with those of Helminthosporiose.
7. The pathogen *Cercospora eleusine* may survive between millet crops in and on infected crop residue as conidia and myacelial masses.

Answer.

S. No	Answer	S. No	Answer
1	True	5	True
2	True	6	True
3	True	7	True
4	True	8	-

B. Descriptive Questions

a. Long answer questions

1. Briefly describe the causal organism, diagnostic symptoms and disease management of the following finger millet disease.

   a. Blast

   b. Leaf spot

2. Describe in detail the symptoms, causal organism, disease-cycle and management of Leaf spot of finger millet.
3. Write the causal organism and their etiology , symptoms and management of following finger millet disease:

   a. Blast

   b. Leaf spot

4. Describe the symptoms and disease management of any two plant diseases caused by the following pathogens:

   (i) *Magnaporthe grisea*

   (ii) *Cercospora eleusine*

5. Describe the blast disease of Finger millet under heads, diagnostic symptoms, causal organism, disease cycle and disease management.

6. Write down the major fungal diseses of fingermillet? Describe any one in detail under heads, diagnostic symptoms, causal organism and management.

b. Short answer questions

1. How would you identify the finger millet disease in the field
   (a) Blast
   (b) Leaf spot
2. Write the epidemiological condition for development of leaf spot finger millet.
3. Write short notes on integrated management of leaf spot finger millet.
4. Write short notes on etiology of blast disease finger millet .
5. Write down the systemic position of pthogen *Magnaporthe grisea.*

Fig. 1: Blast

Fig. 2: Leaf spot

Plate 6: Photograph showing symptoms of major diseases of finger millet

7

Ground Nut Crop Diseases& Management

Major Diseases	Causal Organism
Early leaf spot	*Cercopora arachidicola*
Late leaf spot	*Phaeoisariopsis personata*
Rust	*Puccinia arachidis*
Wilt	*Fusarium oxysporum* and *F. solani*
Collar rot or Seedling blight or Crown rot	*Aspergillus niger and A. pulverulentum*
Root rot	*Macrophomina phaseolina*
Anthracnose	*Colletotrichum dematium and C. capsici*
Bud necrosis	*Peanut bud necrosis virus*
Kalahasti malady	*Tylenchorhynchus brevilineatus*

Tikka or Leaf Spot

Cercospora arachidicola/ Phaeoisariopsis personatum

Diagnostic Symptoms

- Disease is caused by two different pathogens, namely, *Cercospora arachidicola and Phaeoisariopsis personatum.*
- The two pathogens mostly occur on the same leaves.
- The *Cercospora arachidicola* appears about a month earlier than the *Phaeoisariopsis personatum.*
- The two pathogens cause symptoms differing each other.

Early leaf spot

- Sub-circular dark brown spots having 1 to over 10 mm diameter are produced on the upper leaflet surface.
- Spots are of lighter shade of brown on the lower side of the leaflets.
- Yellow halo is seen around the brown spots.
- Dark brown in colour on the upper leaflet surface where most sporulation occurs and a lighter shade of brown on the lower leaflet surface.

- Distribution of fruiting structures is random on upper leaflet surface.
- Oval to elongate spots are also seen on stems, petioles and pegs.
- These are oval to elongate in shape and have more distinct margins than the leaflet lesions.
- On severity of the diseases, leaflets become chlorotic, then necrotic, lesions coalesce and leaflets are shed(Fig-2).

Late leaf spot

- Late leaf spots are nearly circular and darker than early leaf spots.
- Lesions circular to sub-circular in shape with 1 to 6 mm diameter.
- All lesions are dark brown to black in colour.
- On the lower leaflet surface where most sporulation occurs the lesions are black in colour and slightly rough in appearance.
- Fruiting structures are in concentric rings on the lower leaflet surface.
- Lesions on other parts and effects on disease severity are similar to that of early leaf spot(Fig-1).

Table Comparison of early and late leaf spots

Characteristics	Early Leaf Spot	Late Leaf Spot
Causal organism	*Cercospora arachidicola*	*Cercospora personata/ Phaeoisariopsis personatum*
Sexual stage	*Mycosphaerella arachidis*	*Mycosphaerella berkeleii*
Stage of Occurrence	Early infection	Late infection
Shape of spot	Circular to irregular	Usually circular
Yellow halo	Spots are surrounded by yellow halo Yellow halo is indistinct or not present on the lower leaf surface	No yellow halo around spots in the early stages of development
Surface on which	Upper	Lower
Colour of spot on upper leaf surface	Light brown to black tending toward brown	Brown to black tending toward black
Colour of spot on lower leaf surface	Brown	Black
Conidia	Mostly confined to the upper leaf surface but are occasionally found on the lower leaf surface	Remain restricted to the lower leaf surface and cushion of conidiophores can be seen in concentric rings
Most spore are produced on their arrangement	Random	In concentric rings

Etiology

Scientific Classification

Kingdom : Fungi

Division : Eumycota

Sub- division : Deuteromycotina

Class : Hyphomycetes

Order : Hyphomycetales (Moniliales)

Family : Dematiaceae

Genus : *Cercospora/Phaeoisariopsis*

Species : *arachidicola/personatum*

Character	*Cercospora arachidicola*	*Phaeoisariopsis personatum*
Mycelium	Primarily intercellular but it becomes intracellular in later stages when the host cells die	Septate, entirely internal,intercellular
Haustoria	No haustoria are produced	Haustoria produced
Conidiophores	Light olivaceous brown, unbranched, geniculate, septate, and are produced on small stromata.	Uniformally olivaceous brown, small unbranched , septate and geniculate (knee shaped),develop on a stroma emerge in groups by rupturing the epidermis
Conidia	Sub hyaline to yellow brown in color, obclavato-cylindric in shape, 35 X 110-2.5 X5.4μm in size and septate (4-12 septa)	Obclavate to obclavato cylindric, colored, 18 X 60-6 X 10μm in size spetate (1-7 septa)
Perfect stage produces	Perithecia and Ascostromata	Perithecia and Ascostromata
No of ascospore produced	08	08
Ascospore	Hyaline, slightly curved, 2 celled, apical cell larger than lower cell	2 celled and constricted at septum and hyaline

Disease Cycle

- Pathogen survives for long period of time in the infected plant debris through conidia, dormant mycelium and perithecia in soil.
- Volunteer groundnut plants also harbour the pathogen.
- Primary infection is by ascospores or conidia from infected plant debris or infected seeds.

- Secondary spread is by wind borne conidia.
- Rain splash also helps in the spread of conidia.
- Conidia germinate by germ tubes and cause infection directly or through stomata.
- Incubation period is 8-15 days.

Epidemiology

- Warm and moist weather.
- Optimum temperature - 24-28°C.
- High relative humidity- 90% for a period of 3 days.
- Heavy doses of nitrogen and phosphorus fertilizers.
- Deficiency of magnesium in soil.

Management

- Grow tolerant varieties like ICGV 89104, ICGV 91114 (EM), ICGV 920920, ICGV 92093 (MM).
- Seed dressing with Benlate and Vitavax (2 gm/kg of seed).
- Intercropping pearl millet or sorghum with groundnut (1: 3) .
- Follow cereal-cereal- groundnut crop rotation.
- Deep burying of crop residues in the soil, and removal of volunteer groundnut plants .
- Foliar application of aqueous Neem leaf extract (2-5%) or 5% Neem seed kernel extract at 2 week's interval 3 times starting from 4 weeks after planting.
- Foliage spray with Bordeaux Mixture (4:4:50), Mancozeb (0.2%) or Carbendazim (0.1%) if required repeat after 15 days.

Rust

Diagnostic Symptoms

- Every aerial portion of the plant is affected by the disease. Usually, the disease is found when the plants are roughly six weeks old.
- On the underside of leaves, tiny, dusty, brown to chestnut-colored pustules called uredosori develop.

- A powdery mass of uredospores is visible when the epidermis ruptures.
- The upper surface of leaves develops tiny, necrotic dark patches that correspond to the sori.
- On the petioles and stem, rust pustules are visible.
- Brown teliosori, which resemble dark pustules, emerge among the necrotic areas towards the end of the season.
- Lower leaves that are severely infected dry out and drop early. Small, shriveled seeds are produced as a result of the serious disease(Fig-3).

Etiology

- Pathogen: *Puccinia arachidis.*
- There are uredial and telial phases produced by the pathogen.
- In groundnuts, uredial stages are produced in large quantities, but telia production is restricted.
- Uredospores are yellow, circular, oval, or pedicellate, unicellular, and echinulated with two or three germpores.
- Teliospores have two cells and a dark brown color.
- No records exist for pycnial or aecial stages, and knowledge regarding the function of an alternate host is nonexistent.

Disease cycle

- On volunteer groundnut plants, the pathogen survives as uredospores.
- The fungus can also thrive in soil that has been infected by plant debris.
- Uredospores are mostly dispersed via wind-borne inoculum.
- Additionally, the uredospores propagate through contaminating seeds and pods. Dissemination aids include rainsplash and tools.
- Additionally, the fungus thrives on secondary hosts such as *Arachis marginata*, *A. nambyquarae*, and *A. prostrate.*

Epidemiology

- High relative humidity (above 85 per cent).
- Heavy rainfall.
- Low temperature (20-25°C).

Management

- Avoid monoculturing of groundnut.
- Remove volunteer groundnut plants and reservoir hosts.
- Spray Mancozeb (0.25%) Wettable Sulphur (0.3 %) or Tridemorph (0.1%) or Chlorothalonil (0.2%).
- Grow moderatelyresistant varieties like ALR 1.

Wilt

Fusarium oxysporum

Diagnostic Symptoms

- Germinating seeds are attacked by the pathogens shortly before emergence.
- General tissue disintegration and the surface of the seedling is covered with sporulating mycelium.
- Damping off symptoms characterized by brown to dark brown water soaked sunken lesions on the hypocotyl which later encircle the stem and extend above the soil level.
- Roots are also attacked, especially the apical portions.
- Affected seedlings become yellow and wilted.
- Leaves turn greyish green and the plants dry up and die.
- Roots and stems show internal vascular browning and discoloration.
- Pathogens are also commonly associated with pod rot(Fig-4).

Etiology

Scientific Classification

Kingdom : Fungi

Division : Eumycota

Sub- division : Deuteromycotina

Class : Hyphomycetes

Order : Hyphomycetales (Moniliales)

Family : Dematiaceae

Genus : *Fusarium*

Species : *oxysporum*

- Aerial mycelium first appears white, and then may change to a variety of colors.
- Pathogen produces three types of asexual spores; macroconidia, microconidia and chlamydospores.
- Macroconidia are nearly straight, slender and thin-walled. They usually have three or four speta, a foot-shaped basal cell and a curved and tapered apical cell. They are generally produced from phialides in false heads by basipetal division.
- Microconidia are ellipsoidal and have either a single septum or none at all.
- They are formed from phialides in false heads by basipetal division.
- Chlamydospores are globose with thick walls. They are either formed from hyphae or by the modification of hyphal cells.

Disease Cycle

- Pathogen survives in the soil between crop cycles in infected plant debris.
- The pathogen can survive either as mycelium, or as any of its three different spore types *viz*., macroconidia, microconidia and chlamydospores.
- Pathogen can invade a plant either with its sporangial germ tube or mycelium by invading the plant's roots.
- Roots can be infected directly through the root tips, through wounds in the roots, or at the formation point of lateral roots, Once inside the plant, the mycelium grows through the root cortex intercellulary.
- When the mycelium reaches the xylem, it invades the vessels through the xylem's pits. At this point, the mycelium remains in the vessels, where it usually advances upwards toward the stem and crown of the plant.
- As it grows, the mycelium branches and produces microconidia, which are carried upward within the vessel by way of the plant's sap stream. When the microconidia germinate, the mycelium can penetrate the upper

wall of the xylem vessel, enabling more microconidia to be produced in the next vessel.

- The plant's water supply is significantly impacted by the pathogen's development within its vascular tissue.
- The plant eventually perishes as a result of the leaves' stomata closing due to the lack of water.

Epidemiology

- Warm and moist weather.
- Optimum temperature - 24-28°C.

Management

- Seed treatment with Carbendazim 2g/kg seed or Captan 3g/kg seed or Thiram @ 4g/kg seed.
- Avoid contamination of clean fields with soil or crop debris from infested fields.
- Adequate irrigation should be applied.
- Harvest early where a large number of plants are infected early in the season.
- Avoid crop rotations with susceptible crops such as cotton and potato.
- Long rotations with non-hosts such as corn, grain sorghum, and sudan grass.
- Seed treatment with spores and mycellial fragments of *Trichoderma viride* .
- Spot drench with Carbendazim at 0.5 g/lit, if required repeat after fifteen day.

Model Question Paper

A.Objective Type Questions

a.Choose the correct answer from the following

(1) *Phaeoisariopsi personatum* belongs to order.

a. Agonomycetales   b. Peronosporales

c. Hyphomycetales   d. Moniliales

(2) Early leaf spot of groundnut is caused by :

a. *Cercospora arachidicola*　　b. *Rhizoctonia bataticola*

c. *Sphacelotheca sorghi*　　d. *Sphacelotheca cruenta*

(3) Late leaf spot of groundnut is caused by

a. *Cercospora arachidicola*　　b. *Phaeoisariopsis personatum*

c. *Sphacelotheca sorghi*　　d. *Sphacelotheca cruenta*

(4) The symptom of which disease remain restricted to the lower leaf surface and cushion of conidiophores can be seen in concentric ringsis

a. Early leaf spot　　b. Late leaf spot

c. Wilt　　d. None of these

(5) Secondary spread of leaf spot disease of groundnut is by.

a. Soil borne conidia　　b. Air borne conidia

c. Externally seed borne conidia　　d. Internally seed borne conidia

(6) Spots are surrounded by yellow halo is characterisitic symptom of which disease of groundnut

a. Early leaf spot　　b. Late leaf spot

c. Wilt　　d. None of these

(7) The teleomorphic stage of *Cercospora arachidicola*is

a. *Mycosphaerella arachidis*　　b. *Sphacelotheca cruenta*

c. *Rhizoctonia bataticola*　　d. None of these

(8) The teleomorphic stage of *Phaeoisariopsis personatum*

a. *Mycosphaerella arachidis*　　b. *Mycosphaerella berkeleii*

c. *Rhizoctonia bataticola*　　d. None of these

Answer

Sl.No	Answer	Sl.No	Answer
1	c.Hyphomycetales	5	b.Air borne conidia
2	a.*Cercospora arachidicola*	6	a. Early leaf spot
3	b.*Phaeoisariopsis personatum*	7	a.*Mycosphaerella arachidis*
4	b.Late leaf spot	8	b.*Mycosphaerella berkeleii*

b. State whether the following statements are *True* or *False*

1. Leaf spot pathogen of groundnut survives for long period of time in the infected plant debris through conidia, dormant mycelium and perithecia in soil.
2. Primary infection of leaf spot disease in groundnut takes place by ascospores or conidia from infected plant debris or infected seeds.
3. Secondary spread of leaf spot disease of groundnut is by wind borne conidia.
4. The pathogen *Fusarium oxysporum* produces three types of asexual spores; macroconidia, microconidia and chlamydospores.
5. The plant's water supply is greatly affected due to the growth of the pathogen *Fusrium oxysporum* within the groundnut vascular tissue.
6. The *Cercospora arachidicola* appears about a month earlier than the *Phaeoisariopsis personatum* in groundnut.
7. Yellow halo around spots are not seen in the early stages of development of late leaf spot in groundnut.
8. Spots are surrounded by yellow halo in early leaf spot of groundnut.
9. Conidia of *Phaeoisariopsis personatum* remain restricted to the lower leaf surface of groundnut and cushion of conidiophores can be seen in concentric rings.
10. *Phaeoisariopsis personatum* takes nutrition from host through haustoria.

Answer

S. No	Answer	S. No	Answer
1	True	6	True
2	True	7	True
3	True	8	True
4	True	9	True
5	True	10	True

B. Descriptive Questions

a. Long answer questions

1. Describe in detail the symptoms, causal organism, etiology, disease-cycle, epidemiology and management of wilt of groundnut.

2. Describe in detail the symptoms, causal organism, etiology, disease-cycle, epidemiology and management of rust of groundnut.
3. Describe the late leaf spot of ground nut under heads, diagnostic symptoms, causal organism, disease cycle and disease management.
4. Which are the important fungal diseases of groundnut ? Describe any one under heads, diagnostic symptoms, causal organism, disease cycle and disease management.
5. Write short notes on
   a. Disese cycle of wilt of groundnut.
   b. Management of rust of groundnut.
   c. Etiology of *Cercopora arachidicola* cusing early leaf spot of groundnut.

b. Short answer questions

1. Differentiate between early and late leaf spot of groundnut.
2. How would you identify the flowing disease in the field : (a) wilt of groundnut(b) rust of groundnut.
3. Write the epidemiological condition for wilt of groundnut.
4. Write down the etiology of *Phaeoisariopsis personata.*
5. Which fungal pathogen cause early leaf spot of groundnut? Give its systematic position.
6. Elaborate disease cycle of late leaf spot of groundnut in pictorial form .
7. Describe in brief bout management of leaf spot of groundnut.

c. Very short answer questions

1. Name the pathogen or causal organism of following groundnut disease;

   (i) Late leaf spot (ii) Early leaf spot

   (iii) Rust (iv) Wilt
2. Write short notes on management practices of wilt of groundnut.
3. Discuss epidemiology leaf spot of groundnut.
4. Differentiate between wilt and root rot.
5. Name the three fungal diseases of groundnut with their causal organism.

Fig. 1: Late leaf spot

Fig. 2: Early leaf spot

Fig. 3: Rust

Fig. 4: Fusarium wilt

Plate 7: Photograph showing symptoms of major diseases of groundnut

8

Soybean Crop Diseases & Management

Major Diseases	Causal Organism
Rhizoctonia aerial blight / Web blight	*Rhizoctonia solani*
Charcoal rot	*Rhizoctonia bataticola*
Seed and Seedling rot	*Phytophthora, Rhizoctonia, Pythium, Fusarium*
Mosaic	Soybean mosaic virus (SMV)
Bacterial pustule	*Xanthomonas axonopodis* pv. *glycines*
Bacterial blight	*Pseudomonas savastanoi* pv. *glycinea*
Wilt	*Fusarium* oxysporum f. sp. *tracheiphilum*
Frog eye Leaf spot	*Cercospora sojana*
Rust	*Phakopsora pachyrhizi* and *Phakopsora meibomiae*
Anthracnose	*Colletotrichum truncatum* and *Glomerella glycines*
Downy mildew	*Peronospora manshurica*

1. Rhizoctonia aerial blight / Web blight

Symptoms

- The soybean Rhizoctonia aerial blight is a damaging foliar disease that quickly defoliates plants in warm, humid climates.
- *Rhizoctonia solani* causes soybean aerial blight, which manifests as leaf and pod spots, leaf blight, defoliation, lesions on the stem and petiole, a mycelium that resembles a cobweb, and sclerotia that form on diseased leaves.
- The plant's lower or central sections are where infection starts, and it progressively spreads upward.
- The leaves first have wet patches that eventually become green to reddish brown in color before turning tan, brown, or black.
- Small spots at first eventually combine to cover a significant portion of the leaf.
- Affected leaves droop and stick to the pods and stem underneath them during the aerial blight phase, infecting the pods and seeds in the process.

- Brownish lesions appear on the petioles and stem.
- Lesions on the pods might appear as little, brownish dots or as a full pod blight. Pod infection and seed infection are related.
- Elevated relative humidity fosters the development of sclerotia and mycelial growth in blighted and lesioned areas.
- The pathogen creates two different types of sclerotia: macro and micro sclerotia, both of which can infect plants that are six weeks old(Fig-1)..

Etiology

Anamorphic stage

Subdivision: Deuteromycotina

Form class: Deuteromycetes

Form subclass: Hyphomycetidae

Form order: Aganomycetales

Genus: *Rhizoctonia*

Species: *solani*

Teleomorphic stage

Subdivision: Basidiomycotina

Class: Basidiomycetes

Sub class: Holobasidiomycetidae

Order: Tulasnellales

Family: Caratobasidiaceae

Genus: *Thanatephorus*

Species: *Cucumeris*

- The pathogen that causes soybean aerial blight is called *Rhizoctonia solani* Kuhn., which is the asexual, imperfect, or anamorphic stage.
- *Thanatephorus cucumeris* (Frank) Donk., which is the sexual, perfect, or teleomorphic stage.
- A worldwide soilborne fungus with a broad host range is *Rhizoctonia solani.*

- Numerous plants and weeds have reportedly been attacked.
- It was initially documented in India in 1967 from Pantnagar, Uttrakhand.
- There is barely a plant that the fungus *R. solani* cannot infect.
- The pathogen has a very broad host range.

Disease Cycle

- *Rhizoctonia solani,* the fungus that causes aerial blight, overwinters as sclerotia in soil or plant remnants from the previous crop.
- The sclerotia germinate and the mycelium spreads inside the soybean plant when the growing season begins and favorable climatic conditions are present.
- The disease's common symptoms, such as the growth of web-like mycelium, are displayed by the affected plants.
- Through direct contact, flood irrigation, and runoff from rainwater, the fungus spreads to nearby plants.
- Following crop harvest, the fungus persists in soil and plant debris as latent mycelium and sclerotia, respectively.

Epidemiology

- High relative humidity.
- Temperature - 25 to 32°C.

Management

- Destruction of diseased crop residues /debris.
- Deep plowing in fields where disease is common may be helpful.
- Rotate with non-host crops.
- Seed treatment with Carbendazim or Thiram at 2 g/kg.
- Pellet the seeds with *Trichoderma viride* at 4 g/kg or *Pseudomonas fluorescens* @ 10g/kg of seed.
- Apply neem cake at 150 kg/ha.

2. Charcoal Rot

Diagnostic Symptoms

- Initially the symptom starts with yellowing and drooping of the leaves.
- Leaves later fall off and the plant dies within week.
- Dark brown lesions are seen on the stem at ground level and bark shows shredding symptom.
- Affected plants can be easily pulled out leaving dried, rotten root portions in the ground.
- Rotten tissues of stem and root contain a large number of black minute sclerotia(Fig-2).

Etiology

Scientific Classification

Kingdom : Fungi

Division : Eumycota

Sub- division : Deuteromycotina

Class : Hyphomycetes

Order : Agonomycetales

Family : Agonomycetaceae

Genus : *Rhizoctonia*

Species : *bataticola*

- Pathogen: *Macrophomina phaseolina.*
- Pathogen produces dark brown, septate mycelium with constrictions at hyphal branches.
- Minute, dark, round sclerotia produce in abundance.
- Pathogen also produces dark brown, globose ostiolated pycnidia on the host tissues.
- Pycnidiospores are thin walled, hyaline, single celled and elliptical.

Disease Cycle

- Pathogen overwinters as jet black, round to oblong or irregular microsclerotia and resting mycelium in dry soils and embedded in plant residues.
- Primary spread is through seed-borne and soil-borne sclerotia.
- Pathogen restricts water movement in the plant by mechanical plugging of the water-conducting vessels with mycelium and microsclerotia and by secreting toxins and enzymes that kill host tissues.
- Secondary spreads is through pycnidiospores which are air-borne.

Epidemiology

- Day temperature -30°C.
- Prolonged dry season followed by irrigation.
- High soil moisture - but not saturated.
- Soil types with high amounts of organic matter.
- Delayed emergence.

Management

- Seed treatment with Carbendazim or Thiram at 2 g/kg or
- Pellet the seeds with *Trichoderma viride* at 4 g/kg or *Pseudonomas fluorescens* @ 10g/kg of seed.
- Apply neem cake at 150 kg/ha.
- Follow deep ploughing.

3. Seed and Seedling Rot

Causal organsims

Disease	Causal Organisms
Seed rot	*Pythium, Phytophthora, Phomopsis*
Seedling mortality (Damping off, Seedling blight)	*Phytophthora, Rhizoctonia, Pythium*
Root and lower stem decay	*Rhizoctonia, Fusarium, Phytophthora*

Diagnostic Symptoms

Disease	Diagnostic Symptoms
Seed rot	• Soft decay of seed • Missing seedlings in row
Seedling mortality (Damping off, Seedling blight)	• Wilting • Yellowing of leaves • Necrotic lesions on stems • Quick death of seedlings • Leaves remain attached to stem(Fig-3)
Root and lower stem decay	• Reddish-brown lesions on taproot and hypocotyl often superficial. • Phytophthora causes brown lesions on stem above soil-line

Disease Cycle

- Pathogens that attack soybean seeds and seedlings (Phytophthora, Pythium, Rhizoctonia and Fusarium) survive in diseased plant material and in the soil.
- These diseases are most common when soil is very wet in the first few weeks after planting, especially in heavy, poorly drained, compacted or high-residue fields.

Epidemiology

- Pythium and Fusarium are more likely to occur when soil temperatures are cooler (<15°C).
- Phytophthora and Rhizoctonia are more likely to be the culprit if soils are warmer (21°C to 26°C).

Pythium

- Pathogen may attack seeds before or after germination; seeds killed before germination are soft and rotted with soil adhering to them.
- Emerged plants may be killed before the first true leaf stage. These plants have a rotted appearance.
- Diseased plants may easily be pulled from the soil because of rotted roots.
- Plants may be killed by "damping off" before or after emergence. On infected plants, the hypocotyl becomes narrow and is commonly "pinched off" by the disease.

- Prefers cold soil temperatures (<60°F); may be the first soybean disease found in a growing season.
- High-residue fields, and heavy or compacted soils are at higher risk because of cooler, wetter conditions.

Rhizoctonia

- Infection is characterized by a shrunken, reddish-brown lesion on the hypocotyl at or near the soil line.
- Infection may be superficial, causing no noticeable damage, or may girdle the stem and kill or stunt plants.
- Normally appears as the weather becomes warm (~80F); more often seen in late-planted soybean fields.
- Causes loss of seedlings (damping-off) in small patches or within rows; is usually restricted to the seedling stage
- Is more common in wet soils or moderately wet soils where germination is slow or emergence is delayed.

Fusarium

- Causes light- to dark-brown lesions on soybean roots that may spread over much of the root system.
- May attack the tap root and promote adventitious root growth near the soil surface, and may also degrade lateral roots, but usually does not cause seed rot.
- Infection is caused by a complex of different species that prefer different conditions; some prefer warm and dry soils, while others prefer cool and wet soils.

Management

- Grow disease resistant cultivars like PS 16 and Pusa Bisaki.
- Seed treatment with 4g *Trichoderma viride* formulation or with 3g Thiram per kilogram of seed.

Bacterial Pustule

Diagnostic Symptoms

- Most prevalent later in the growing season.

- Initial symptoms are the appearance of tiny pale green spots on leaves. These spots have raised centers that may develop on either surface of the leaf but are more common on the lower surface.
- Lesions are often associated with main leaf veins .
- As the disease progresses, small, light-colored pustules will form in the center of the spots.
- Spots may merge together to form irregular lesions.
- Symptoms of bacterial pustule are very similar to those of bacterial blight during the early stages of development, but water soaking will not be present with bacterial pustule, and bacterial pustule lesions will have raised centers as they develop(Fig-6).

Etiology

Scientific Classification

Kingdom : Prokaryotae

Division : Gracilicutes

Class : Proteobacteria

Class : Xanthomonadales

Family : Xanthomonadaceae

Genus : *Xanthomonas*

Species : *axonopodis* pv. *glycines*

- Pathogen: *Xanthomonas axonopodis* pv. *glycines*
- Cells appeared as single, some in pairs and in clusters having single polar flagellum.
- Cells measured 0.4 -0.25 x 1.25- 3.00 mm in size.
- Circular, flattened, slightly raised, convex with yellow to bright yellow colour on SX and NA medium.
- They occurred singly or rarely in aggregates.
- Bacterium liquified the gelatin, produced the H_2S, positive for starch hydrolysis and acid produced by carbon sources (sucrose,glucose and dextrose) but it was negative for acid production by nitrogen (peptone) sources.

Disease Cycle

- Pathogen overwinters in crop residue.
- Carried by wind-driven rain or water droplets splashing from the ground to the plant.
- Disease can be spread during cultivation while the foliage is wet.
- Bacterium will enter the plant through natural openings and wounds.

Epidemiology

- Warm weather with frequent showers.
- Optimum temperature- 86-92^0 F.

Management

- Use of disease free and clean seed harvested form healthy plants.
- Use of disease resistant varieties *vix,* PK332, PK416, PK1029,PK1042,JS 71-05, JS90-41, Himso 1563, Indira Soya-9, KHSb 2, MAUS-32, NRC-7,NRC-37,SL-96, VLS-2
- Seed treatment with Streptocycline @500 mg/kg seed.
- Foliar spray of Copper oxychloride (0.3%) + Streptocycline (0.01%) at the time of disease appearance and if required repeat after 15 days of first spray.
- Destruction of diseased crop residues /debris.
- Deep plowing in fields where disease is common may be helpful.
- Rotate with non-host crops.

Bacterial Blight

Pseudomonas savastanoi pv. *glycinea*

Diagnostic Symptoms

- Typically an early season disease.
- Plants infected early in the growing season are characterized by brown spots on the margins of the cotyledons.
- Young plants may be stunted, and if the infection reaches the growing point, they may die.

- Symptoms in later growth stages include angular lesions, which begin as small yellow to brown spots on the leaves(Fig-5) .
- Centers of the spots will turn a dark reddish-brown to black and dry out. A yellowish-green "halo" will appear around the edge of the water-soaked tissue that surrounds the lesions.
- Eventually, the lesions will fall out of the leaf and the foliage will appear ragged.
- Generally, young leaves are most susceptible to blight infection.
- The disease rarely affects seeds, but when lesions do appear on pods, developing seeds may become shriveled and discolored.

Disease Cycle

- Pathogen overwinters in the field in plant residue.
- Seedlings may be infected through infested seed.
- Initial infection of soybeans occurs when bacterial cells are carried by splashing or wind-driven water droplets from plant residue on the soil surface to the leaves.
- Disease outbreaksusually follow rainstorms with high winds.
- Bacteria enter the plants through stomata and wounds on leaves.
- In order for infection to occur, the leaf surface must be wet.

Epidemiology

- Cool, rainy weather (70-80^{0}F).
- Wind driven rains.
- Continuous soybean fields.
- No-till soybean fields.

Management

- Plant disease-free seed.
- Use disease resistance varieties like Bragg, PK262, PK416, PK472, PS1024, PS1042
- Foliar spray of Copper based fungicide (0.3%) and Streptocycline (0.01%) at the time of disease appearance.

- Avoid highly susceptible varieties in areas where bacterial blight is serious.
- Rotate crops with at least one year between soybean crops.
- Do not cultivate when foliage is wet.

Table 1: Comparison of bacterial disease characteristics of soybean.

Characteristics	Bacterial Blight	Bacterial Pustule
Overwinter	Crop residue	Crop residue
Initial Symptoms	Yellow-brown spots with water soaked underside of leaf lesion	Tiny pale green spots on leaves
Later Symptoms	Irregularly shaped lesions with water-soaked tissue and yellow halo	Circular spots with elevated centers
Optimum Temperature	70-80°F	86-92°F

Mosaic

Soybean mosaic virus (SMV)

Diagnostic Symptoms

- Disease is occurring all over the soybean growing areas of India and reduces yield ranging from 50 to 93 percent.
- Diseased plants are usually stunted with distorted (puckered, crinkled, ruffled, narrow) leaves.
- Pods become fewer and smaller seeds. Infected seeds get mottled and deformed. Infected seeds fail to germinate or they produce diseased seedlings(Fig-4).

Etiology

- Caused by *Soybean mosaic virus* - a potyvirus.
- Virus is flexuous rod belongs to potyvirus group particles 750 -900nm long.
- Genome- ss RNA.

Disease Cycle

- Virus is seed, graft and sap transmitted.
- Narrow host range, transmitted in a non persistent manners by about 32 species of aphid belonging to 15 different genera.
- Extent of damage to crop depends on age of plant at the time of infection.

Epidemiology

- Optimum temperature -around 18^0 C.
- Symptoms are masked above 30^0C.
- Humid weather.

Table Characteristics of other major viruses associated with soybeans

Virus	Symptoms	Host	Transmission
Mung bean yellow mosaic virus	Dark yellow colored spot on leaves. Yellow area is either scattered produced indefinite bands along with major areas.	Soybean, Mungbean	Transmitted by white fly (*Bemisia tabaci*)
Bean pod mottle virus (BPMV)	Young leaves in the upper canopy exhibit light green to yellow mottling; some puckering and distortion; stems remain green with mature pods; retain petioles after leaf blades drop. Discolored seed from pigments that bleed from hilum.	Soybean Snap bean	Bean leaf beetle is Common vector; seed transmission is less than 1%.
Tobacco streak virus (TSV)	Symptoms are mild mosaic of yellow and green; leaf and floral buds may proliferate excessively; pods mature but stems remain green	Soybean Alfalfa Clovers Tobacco Snap bean	Pollen transmitted; Thrips are reported as vector. Seed transmission as high as 50%.
Bean yellow mosaic virus (BYMV	Bright yellow mottling of leaves; crinkled leaves; yellow areas are either scattered or produced in indefinite bands; rusty, necrotic spots appear in yellow areas as leaves mature; leaf veins may turn dark; BYMV will cause delayed maturity	Soybean Snap bean Pea Clovers	20 species of aphids reported to transmit BYMV; seed transmission is not reported.
Alfalfa mosaic virus (AMV)	Bright yellow mosaic of leaves; or leaf veins are yellow but remainder of leaf is a normal green color.	Soybean Alfalfa Clovers Snap bean Pea	Transmitted by aphids; seed transmission is low, 1-5%

Management of Viruses

- Deep summer ploughing.
- Use healthy/certified seeds.
- Use of moderately resistant varieties like JS 71-05, KHSb-2,LSb-1, MACS58,MACS 124, PK416, Punjab-1.
- Cultivate 2 to 3 or more varieties and change of varieties at least after every two years.

- Rogue out infected plants and burn them.
- Pre-sowing soil application of Phorate @ 10 kg/ha.
- Two foliar sprays of Thiamethoxam 25 WG @ 100 g/ha at 30 and 45 days after sowing.
- Keep the field free from weeds.

Model Question Paper

A. Objective Type Questions

a. Choose the correct answer from the following :

(1) *Rhizoctonia solani* belongs to order.

a. Agonomycetales b. Peronosporales

c. Hyphomycetales d. Moniliales

(2) Rhizoctonia blight of soybean is caused by

a. *Rhizoctonia solani* b. *Rhizoctonia bataticola*

c. *Sphacelotheca sorghi* d. *Sphacelotheca cruenta*

(3) The teleomorphic stage of *Rhizoctonia solani* is

a. *Thanatephorus cucumeris* b. *Sphacelotheca cruenta*

c. *Rhizoctonia bataticola* d. None of these

(4) The secondary spreads of *Rhizoctonia bataticola* in soybean is through pycnidiospores which are

a. Soil and seed borne b. Air borne

c. Externally seed borne d. Internally seed borne

(5) *Soybean mosaic virus* is transmitted by

a. White fly b. Aphid

c. Leaf hopper d. None

(6) The diagnostic symptom like water soaking will not be present and lesions will have raised centers as they develop is related to which bacterial disease of soybean

a. Bacterial blight b. Bacteiral pustule

c. Bacterial leaf spot d. None

(7) Mung bean yellow mosaic virus of soybean is transmitted by :

a. White fly b. Aphid

c. Leaf hopper d. None

(8) Bean pod mottle virus of soybean is transmitted by

a. White fly b. Aphid

c. Leaf hopper d. Bean leaf beetle

(9) Tobacco streak virus of soybean transmitted by

a. White fly b. Aphid

c. Leaf hopper d. Thrips

(10) Bean yellow mosaic virus of soybean is transmitted by

a. White fly b. Aphid

c. Leaf hopper d. Thrips

a. Answer

S. No	Answer	S. No	Answer
1	a. Agonomycetales	6	b.Bacteiral pustule
2	a. *Rhizoctonia solani*	7	a.White fly
3	a. *Thanatephorus cucumeris*	8	a.White fly
4	b. Air borne	9	d.Thrips
5	a. White fly	10	b.Aphid

b State whether the following statements are *True* or *False*

1. The pathogen *Rhizoctonia bataticola* produces dark brown, globose ostiolated pycnidia on the soybean host tissues.
2. The pathogen *Rhizoctonia bataticola* overwinters as jet black, round to oblong or irregular microsclerotia and resting mycelium in dry soils and embedded in plant residues.
3. The pathogen *Rhizoctonia bataticola* does not produce spores, hence is identified only from mycelia characteristics or DNA analysis.
4. The pathogen *Rhizoctonia solani* is subdivided into anastomosis groups based on hyphal fusion between compatible strains.
5. Seed rot and seedling mortality diseases are most common in soybean when soil is very wet.

6. Foliar sprays of Thiamethoxam 25 WG @ 100 g/ha at 30 and 45 days after sowing mange yellow mosaic of soybean.
7. Bacterial Blight is typically an early season disease of soybean.
8. Cultivation of two or more varieties an an area and change of varieties at least after every two years are found effective in management of yellow mosaic disease of soybean.
9. Soybean mosaic virus is a potyvirus.
10. Foliar spray of Copper based fungicide (0.3%) and Streptocycline (0.01%) at the time of disease appearance control bacterial blight of soybean.

b. Answer

S. No	Answer	S. No	Answer
1	True	7	True
2	True	8	True
3	True	9	True
4	True	10	True
5	True	11	
6	True	12	

B. Descriptive Questions

a. Long answer questions

1. Describe the Rhizoctonia blight of soybean under heads, diagnostic symptoms, causal organism, disease cycle and disease management.
2. Which are the important fungal diseases of soybean ? Describe any one under heads, diagnostic symptoms, causal organism, disease cycle and disease management.
3. Write short notes on

   a) Disese cycle of rhizoctonia blight of soybean

   b) Management of mosaic disease of soybean

   c) Etiology of *Rhizoctonia solani* cusing Aerial blight of soybean.
4. Briefly describe the causl organism, diagnostic symptoms and management of the following plant disease.

   a. Seed and seedling rot of soybean.

   b Bacterial spot of soybean.

5. Describe in detail the symptoms, causal organism, mode of transmission and management of mosaic of soybean.
6. Describe in detail the symptoms, causal organism, etiology, disease-cycle, epidemiology and management of Seed and Seedling rot of soybean.
7. Describe in detail the symptoms, causal organism, etiology, disease-cycle, epidemiology and management of bacterial spot of soybean.

b. Short answer questions

1. Differentiate between wilt and rot symptom.
2. How would you identify the flowing disease in the field : (a) Rhizoctonia blight of soybean (b) Mosaic disease of soybean.
3. Write the epidemiological condition for Rhizoctonia blight of soybean.
4. What do you mean by Aerial blight?
5. Write down the etiology of *Rhizoctonia solani.*
6. Which fungal pathogen cause Charcoal rot of soybean? Give its systematic position.
7. Elaborate disease cycle of rhizoctonia blight of soybean in pictorial form.
8. Describe in brief about mangment of mosaic disease of soybean.
9. Compare between bacterial blight and bacterial pustule of soybean

c. Very short answer questions

1. Name the pathogen or causal organism of following plant disease ; (i) Rhizoctonia blight of soybean (ii) Charcoal rot of soybean (iii) Aerial blight of soybean (iv) Rust of soybean.
2. Write short notes on management practices of Rhizoctonia blight of soybean.
3. Discuss integrated management schedule of Seed and Seedling rot of soybean.
4. Differentiate between blight and spot.
5. Name the three fungal diseases of soybean with their causal organism.

Fig. 1: Rhizoctonia aerial blight

Fig. 2: Charcoal rot

Fig. 3: Seedling disease

Fig. 4: Yellow mosaic

Fig. 5: Bacterial blight

Fig. 6: Bacterial pustule

Plate 8: Photograph showing symptoms of major diseases of soybean

9

Gram Crop Diseases & Management

Name of Diseases	Causal Organism
Ascochyta blight	*Ascochyta rabiei* Pass Labr
Wilt	*Fusarium oxysporum* f.sp. *ciceris* (Padwick)Matuoet. K Sato
Gray mold	*Botrytis cinerea* Pers.Ex.Fr
Rust	*Uromycesciceris-arietini* Grogn.Jacz.&Beyer
Collar rot	*Sclerotium rolfsii* Sacc
Dry root rot	*Rhizoctonia bataticola*Taub Butler
Wet root rot	*Rhizoctonia solani*
Alternaria blight	*Alternaria alternata* (Fr.) Keissl. And *Alternaria tenuis*
Anthracnose	*Colletotrichum dematium*

Ascochyta Blight

Diagnostic Symptoms

- All above ground parts of the plant are infected.
- Round or elongated lesions, bearing irregularly depressed brown spot and surrounded by a brownish red margin on leaf.
- Similar spots may appear on the stem and pods.
- Spots on the stem and pods have pycnidia arranged in concentric circles as minute black dots.
- When the lesions girdle the stem, the portion above the point of attack rapidly dies.
- If the main stem is girdles at the collar region, the whole plant dies (Fig.1).

Etiology

Scientific Classification

Kingdom : Fungi

Division : Eumycota

Sub- division :Deuteromycotina

Class : Coelomycetes

Order :Sphaeropsidales

Family :Sphaeropsidaceae

Genus :*Aschochya*

Species :*rabiei*

- Pathogen: *Ascochyta rabiei*
- Pathogen occurs as both an anamorph (asexual state) and teleomorph (sexual state).
- Anamorph: *Ascochyta rabiei*
- Mycelium: hyaline to brown, septate.
- Pycnidia: fruiting body formed on stems, leaves and pods , round or flattened, globose, dark brown, immersed in the tissues, measuring 100 to 200 µm in diameter, with a prominent ostiole are spherical to sub-globose.
- Conidia: formed inside pycnididum,hyaline, oval to oblong, straight or slightly curved, 1-septate, some 0-septate, slightly or not constricted at the septum, rounded at each end and measure 10-16x3.5 µm in size.
- Teleomorph:*Didymella rabiei(Mycosphaerella rabiei).*
- In perfect stage, the dark colored , globose pseudothecia are formed , each measuring 100 to 140 µm.
- Pseudothecia are characterized by eight ascospore. The ascospores are bicelled with one cell larger than the other,measure 12.5-19.0x6.7 -7.6 µm.

Disease Cycle

- Pathogen survives in the infected plant debris as pycnidia or in seed.
- Primary infection is from inoculums produced by pathogen from seed-borne pycnidia and plant debris in the soil.
- Secondary infection is mainly through air-borne conidia.
- Rain splash also helps in the spread of the disease.

Epidemiology

- Temperature of 20-25°C.
- Relative humidity of 60%.
- High rainfall during flowering.
- Leaf wetness -17h.

Management

- Use only certified seed.
- Use resistant varieties *viz*. Him chana -1, Gaurav, Vardan, Samrat, PBG 1 and BG 261.
- Treat the seeds with Thiram 2g or Carbendazim 2 g or Thiram + Carbendazim (1:1 ratio) at 2 g/kg.
- Exposure of seed at 40-50°C reduced the survival of *A. rabiei*by about 40-70 per cent.
- Foliar spray with Carbendazim (0.1%) or Chlorothalonil (0.15%) solution, if required repeat after 15 days.
- Destroy blight-infested crop residues and volunteer chickpea plants.
- Choose crop rotations that permit 3 to 4 years between successive chickpea crops.

Botrytis Gray Mold

Diagnostic Symptoms

- Disease is usually seen at flowering time when the crop canopy is fully developed.
- Lack of pod setting is the first indication of the disease.
- Under favorable conditions foliage shows clear symptoms and plants often die in patches.
- More severe on portions of the plant hidden under the canopy.
- Shed flowers and leaves, covered with the spore mass, can be seen on the ground under the plants.
- When humidity is very high, the symptoms appear on stems, leaves, flowers and pods as gray or dark brown lesions covered with moldy sporophores.

- Lesions on stem girdle the stem completely.
- Tender branches break off at the point where grey mold has caused rotting.
- Affected leaves and flowers turn into a rotting mass.
- Lesions on the pod are water soaked and irregular.
- Grayish white mycelium may be seen on the infected seeds (Fig.3).

Etiology

- Pathogen: *Boirytis cinerea* pers. Ex. Fr.
- Mycelium: septate, brown, hyphae being 8-16μ m wide.
- Conidiophores: light brown, septate, erect and their tips are slightly enlarged bearing small and pointed sterigmata.
- Conidia: hyaline, 1-celled, oval and are borne in clusters on short sterigmata.Conidia in mass are ash-grey in colour, and measure 4-16 x 4-20μ m.

Disease Cycle

- Pathogen survives in the infected plant debris. However, infected seed is thought to be the primary source of infection.
- Pathogen is a facultative parasite with a very wide host range in the temperate and sub-tropical region.
- It causes gray mould disease in a number of crop plants, such as strawberry, grapevine, apple, cabbage, carrot, cucumber, eggplant, lettuce, pepper, squash, tomato and several ornamentals.
- Pathogen also produces highly resistant sclerotia as survival structures in older cultures. It overwinters as sclerotia or intact mycelia, both of which germinate in spring to produce conidiophores.
- Spores of *B. cinerea* are released from conidiophores and then dispersed by air currents or rain splash and cause new infections.

Epidemiology

- Optimum temperature- 28 to 30 ^{0}C.
- Relative humidity – 95-100 per cent.

Management

- Grow resistant varieties: BG 276, GL 90159, GL 91040, GL 91071, and GL 92162.
- Adopting wider spacing.
- Intercropping with linseed.
- Avoiding excessive vegetative growth.
- Avoiding excessive irrigation.
- Growing compact varieties.
- Seed treatment with Carbendazim + Thiram @ 3 g/kg.
- Foliar spray of Carbendazim @ 1 g/L, if required repeat after 15 days.

Wilt

Diagnostic Symptoms

- At first green chlorosis, typically appear on lower leaves and extending up the plant. Leaves eventually take on a dull-yellow colour, wilt and the plant collapses and dies.
- Adult plants wilt, with their petioles and rachises drooping.In some cases there may be leaf vein clearing before wilt begins.
- Internally, the xylem tissues stain dark brown to almost black.
- Wilting may initially affect only one side of the plant (Fig.2).

Etiology

Scientific Classification

Kingdom : Fungi

Division : Eumycota

Sub- division :Deuteromycotina

Class : Hyphomycetes

Order :Hyphomycetales (Moniliales)

Family :Dematiaceae

Genus :*Fusarium*

Species :*oxysporum*f sp *ciceri*

- Pathogen: *Fusarium oxysporum*f.sp. *ciceris.*
- Mycelium: produces profusely branched septate hyphae, hyaline to light brown in color.
- Macroconidia :1-7 (mostly 3-5) septate, 25-55 to 2.5-6 µm, fusiform, curved, with a tapering, pointed, sometimes hooked, apical cell and distinctly pedicellate basal cell. In some strains macroconidia are sparse, in others, abundantly produced in pinkish sporodochial conidiomata.
- Microconidia: borne on simple short conidiophores arising laterally on the hyphae.These are oval to cylindrical, straight to curved, and 5-15 to 2-5 µm, collecting in small, slimy droplets.
- Chlamydospores: globose, hyaline 7-15 µm diameter, usually forming abundantly in mature colonies, terminal or intercalary, single or in small groups or chains.

Disease Cycle

- Pathogen is mainly soil borne as well as seed borne and also it is a facultative saprophyte.It can survive in the soil upto six years in the absence of susceptible host.
- Primary infection is through chlamydospores in soil, which remain viable upto next crop season.
- Secondary spread is through irrigation water, cultural operations and implements.
- Following infection of host roots, the fungus crosses the cortex and enters the xylem tissues.
- It then spreads rapidly up through the vascular system, becoming systemic in the host tissues, and may directly infect the seed.

Epidemiology

- High soil temperature – 25°C.
- High soil moisture.

Integrated Management

- Grow resistant cultures like Avrodhi, Alok Samrat, Pusa-212, JG- 322 GPF-2, Haryanachana-1 and Kabuli chickpea like Pusa-1073 and Pusa-2024, ICCC 42, H82-2.

- Seed treatment with Carbendazim or Thiram at 2 g/kg or Carbendazim 1 g+Thiram 1g/kgor seeds treatment with *Trichoderma viride*at 4 g/kg or *Pseudomonas fluorescens* @ 10g/kg of seed.
- Apply heavy doses of organic manure or green manure.

Model Question Paper

A. Objective Type Questions

a. Choose the correct answer from the following

(1) *Aschochyta rabiei* belongs to order.

a. Sphaeropsidales b. Peronosporales

c. Hyphomycetales d. Moniliales

(2) Ascochyta blight of gram is caused by

a. *Ascochyta rabiei* b. *Rhizoctonia bataticola*

c. *Sphacelotheca sorghi* d. *Sphacelotheca cruenta*

(3) The teleomorphic stage of *Aschochyta rabiei* is

a.Didymella rabiei b. Sphacelotheca cruenta

c. Rhizoctonia bataticola d. None of these

(4) The teleomorphic stage of *Aschochyta rabiei* form dark colored, golobose

a.Pseudothecia b. Perithecia

c. Apothecia d. Cleistothecia

(5) The fruiting body formed by *Aschochyta rabiei in chickpea is*

a. Pseudothecia b. Pycnidia

c. Apothecia d. Cleistothecia

(6) The secondary spreads of Ascochyta blight in gram is through

a. Soil borne conidia b. Air borne conidia

c. Seed borne conidia d. Vector

(7) The pathogen *Botrytis cinerea* produces highly resistant surviving structure in in old culture called

a. Pseudothecia b. Sclerotia

c. Apothecia d. Cleistothecia

(8) Wilt of chickpea is caused by

a. *Ascochyta rabiei* b. *Rhizoctonia bataticola*

c. *Fusarium oxysporum f.sp. ciceri* d. *Sphacelotheca cruenta*

(9) *Fusarium oxysporum f.sp. ciceri* produces

a. Mcroconidia b. Microconidia

c. Chlaymydospores d. All

(10) Dry root rot of gram is caused by

a. *Rhizoctonia bataticola* b. *Rhizoctonia solani*

c. *Sphacelotheca sorghi* d. *Sphacelotheca cruenta*

(11) Wet root rot of gram is caused by

a. *Rhizoctonia solani* b. *Rhizoctonia bataticola*

c. *Sphacelotheca sorghi* d. *Sphacelotheca cruenta*

(12) Collar rot of gram is caused by :

a. *Rhizoctonia solani* b. *Sclerotium rolfsii*

c. *Sphacelotheca sorghi* d. *Sphacelotheca cruenta*

Answer

S. No	Answer	S. No	Answer
1	a. Sphaeropsidales	7	b. Sclerotia
2	a. *Ascochyta rabiei*	8	c. *Fusarium oxysporum f.sp. ciceri*
3	a. *Didymella rabiei*	9	d. All
4	a. Pseudothecia	10	a. *Rhizoctonia bataticola*
5	b. Pycnidia	11	a. *Rhizoctonia solani*
6	b. Air borne conidia	12	b. *Sclerotium rolfsii*

b Fill in the blanks with suitable words

(1) *Fusarium oxysporum f.* sp. *ciceri* causes .... of chickpea.

(2) Wet root rot of gram is caused by ....

(3) .... is the perfect state or teleomorph of *Ascochyta rabiei.*

(4) Chlamydospore is the resting spore of the fungus ....

(5) Pycnidia is fruiting body formed in ....

S. No	Answer	S. No	Answer
1	Wilt	4	*Fusarium oxysporum f.* sp. *ciceri*
2	*Rhizoctonia solani*	5	*Ascochyta rabiei*
3	*Didymella rabiei*	6	

c State whether the following statements are *True* or *False*

1. The pathogen *Fusarium oxysporum f.sp.ciceri* causing wilt in chickpea is mainly soil borne in nature.
2. *Didymella rabiei* is the perfect state or teleomorph of *Ascochyta rabiei.*
3. Lack of pod setting is the first indication of the botrytis gray mould disease in gram.
4. The fungus *Fusarium oxysporum f.sp.ciceri*produces three types of conidia like macroconidia, microconidia and chlamydospores.
5. Wilt disease of chickpea can be managed by protective application of Trichoderma.
6. Pseudothecia of *Ascochyta rabiei* are characterized by eight ascospore asci.
7. Grayish white mycelium seen on the infected seeds of chickpea is a diagnostic sign of botrytis grey mold disease .
8. The xylem tissues of chickpea infected with wilt disease stain dark brown to almost black.

Answer

S. No	Answer	S. No	Answer
1	True	5	True
2	True	6	True
3	True	7	True
4	True	8	True

Descriptive Questions

a. Long answer questions

1. Describe in detail the symptoms, etiology, disease cycle and management of Ascochyta blight of chickpea.
2. Briefly describe the most distinguishing symptoms and disease management of any two of the following plant disease :

(i) Collar rot of chickpea

(ii) Dry root rot of chickpea

(ii) Wiltof chickpea

3. Describe the wilt of chickpea in the following headings :

(i) Pathogen (ii) Symptoms

(ii) Disease cycle (iv) Disease management

4. How would you identify the following diseases in field

(i) Collar rot of chickpea (ii) Dry root rot of chickpea

(ii) Wilt of chickpea (iv) Wet root rot of chickpea

5. Give the disease cycle and management of the following diseases :

(i) Wilt of chickpea

(ii) Botrytis grey mould of chickpea

6. Describe in detail the most distinguishing symptoms, causal organism, disease cycle and management of botrytis grey mould of chickpea.

b. Short answer questions

1. Compare and contrast wet root rot and dry root rot of chickpea giving suitable measures for their management.
2. Mention on diagnostic symptoms and management of wilt of chickpea.
3. Discuss about the disease cycle and epidemiology of botrytis grey mould of chickpea.
4. Give the disease cycle and predisposition of Ascochyta blight of chickpea.
5. Write down an integrated management practices of Botrytis grey mould of chickpea.

c. Very short answer questions

(1) Suggest some disease resistant varieties of chickpea gainst wilt disease.

(2) Name the diseases caused by the following pathogens :

(a) *Fusarium oxysporum f.* sp. *ciceri*

(b) *Sclerotium rolfsii*

(c) *Uromyces ciceris-arietini*

(d) Rhizoctonia solani

(3) List some important diseases and their causal organisms of gram.

(4) Distinguish between wilt and collar rot of gram.

(5) Name the fungal pathogens causing the following diseases :

(a) Collar rot of gram

(b) Rust of chickpea

(c) Wet root rot of gram

(d) Aschochyta blight of gram

Fig. 1: Ascochyta blight

Fig. 2: Wilt

Fig. 3: Botrytis grey mold

Plate 9: Photograph showing symptoms of major diseases of chickpea

10

Pea Crop Diseases & Management

S. No	Name of Disease	Causal Organism
1	Downy mildew	*Peronospora pisi* Sydow
2	Powdery mildew	*Erysiphe pisi* DC. Syn. *E. polygoni* DC.
3	Rust	*Uromyces fabae (Pers.)* Schrot. &Uromyces *fabae DB. f. sp. pisisativae* Hiratsuka
4	Anthracnose	*Colletotrichum pisi*
5	Fusarium wilt	*Fusarium oxysporum* f. sp. *pisi*

Downy Mildew

Diagnostic Symptoms

- At first grayish white, moldy growth appears on the lower leaf surface, and a yellowish area appears on the opposite side of the leaf.
- Infected leaves can turn yellow and die if weather is cool and damp.
- Stems may be distorted and stunted.
- Brown blotches appear on pods, and mold may grow inside pods(Fig-2).

Etiology

Scientific Classification

Phylum :Heterokontophyta

Class : Oomycota

Order : Peronosporales

Family :Peronosporaceae

Genus :*Peronospora*

Species :*pisi*

- Pathogen: *Peronospora pisi.*
- Pathogen is an obligate parasite.

- Mycelium: coenocytic, hyaline, profusely branched, intercellular sending branched, finger shaped haustoria inside the host cells.
- Conidiophores: unbranched for about two third or more of their length and dichotomously branch at the tip.Branches are long, slender and pointed, and are called sterigmata.A single conidium is produced at the tips of each sterigmata.
- Conidia: oval to elliptical and light yellow in color.Conidia are short lived and germinate readily by germ tube.
- Antheridia &Oogonnia: When the pathogen moves towards the end of the growing season, its hyphae move deep into the host tissue interspaces wherein they produce sex organs (antheridia and oogonia) in close proximity.
- Oospore: After fertilization via fertilization tube, thick walled, green yellow, warty oospores are produced at the rate of one in each oogonium.

Disease Cycle

- Pathogen survives in the soil and an old pea trash through oospores.It can also be seed-borne.
- Oospores perennating in soil germinate by germ tube at the return of favorable conditions.
- The germ tube infects underground parts of young seedlings and the pathogen progresses upwardly systemically causing primary infection..
- These diseased plants provide the wind blown spores that spread the disease in the field and serve as source of secondary infection.
- The oogonial and oospore stages occur later; before harvest, the oospores finally reaches the soil, where they survive one to two years.

Epidemiology

- Temperature- 10-20 ºC.
- Relative humidity- 90%.

Management

- Field sanitation.
- Growing of resistant varieties.
- Seed treatment with Metalaxyl at 2.5g/kg.

- Foliar spray with Metalaxyl + Mancozeb @ 0.2% , if required repeat after 15 days of first spraying..
- Deep ploughing of fields during summer.
- Three summer ploughings at 10 days interval reduces pathogen population.
- Timely sowing should be done.
- Destroy the alternate host plants .
- Apply manures and fertilizers as per soil test.
- Destruction of diseased plant debris.

Powdery Mildew

Diagnostic Symptoms

- At first symptom appears on leaves.
- White floury patches appear on both surface of leaves as well as tendrils, stem and pod.
- As plant become older, the symptoms alsomt cover the entire plant, become more or less grayish brown and the infected parts impart dirty appearance.
- Fruits do not either set or remain very small.
- Later stages, powdery growth also covers the pods(Fig-1).

Etiology

Scientific Classification

Kingdom: Fungi

Division: Eucomycota

Division: Ascomycotina

Class :Pyrenomycetes

Order :Erysiphales

Family :Erysiphaceae

Genus :*Erysiphe*

Species :*polygoni*

- Pathogen: *Erysiphe polygoni*
- Pathogen: obligate parasite and ectophite, only haustoria enter into the cell.
- Mycelium: septate, hyaline, profusely branched, and superficial forming a white web like coating over the surface of the infected part.It sends finger like haustoria into the epidermal cells of the host to obtain nutrients.After maturity, many hyphal branches give rise to erect, stout, conidiophores, which produce conidia in chain.
- Conidiophores:bears spores in chain , septate and their cells appear like conidia in shape.
- Conidia: is elliptical or barrel shaped, hyaline, single celled, and measures 25-35x13-16 μm.
- Cleistothecia: black, minute bodies provided with a number of appendages, and remain scattered in the mycelia web, contains usually 2-8 asci.
- Ascospores: 3-8 , hyaline, one celled, elliptical and unicellular, measuring 19-25x9-14μm.

Disease Cycle

- Pathogen survives in soil and plant debris in the form of cleistotheica and may survive in seed in the form of dormant mycelium.
- Primary infection occurs by asci and ascospores releases after breakdown of cleistotheica or conidia from collateral hosts.
- Secondary spread takes place by wind borne conidia produced as a result of primary infection.

Epidemiology

- Optimum temperature-15-25°C.
- Relative humidity –over 70%.
- Absense of rain.

Management

- Field sanitation.
- Growing of early maturing and resistant varieties like Rachna, Pant P5, DMR 11, HUP 2, JP 885, KFP 103, Ambika, Shubhra, Aparna, Azad P4, Pusa Panna.

- Foliar spray with Carbendazim 50% WP@ 0.1% or Karathane 48% EC @) 0.05% or Wettable Sulphur 40% WP 0.2% or Triadimefon 25% WP @ 0.1% . Second spray after 25 days of interval.
- Control volunteer field peas.
- Avoid sowing field pea crops adjacent to last season's stubble.
- Burn infected pea stubble soon after harvest.

Rust

Diagnostic Symptoms

- Disease appears usually at the beginning of flowering.
- Leaves of infected plants exhibit many small, orange-brown pustules usually at the lower surface.
- Severely infected leaves wither and may drop from the plant.
- Larger pustules occur on the stems and isolated pustules may be found on the pods(Fig-3).

Etiology

Phylum	Basidiomycotina
Class	Teliomyetes
Order	Uridinales
Family	Pucciniaceae
Genus	*Uromyces*
Species	*pisi*

- Pathogen:*Uromyces pisi (Pers.) Schrot.*
- Autoecious rust:Pathogen passes all stages on peas completing its life cycle on pea.
- Aecial stage: appear first on the lower surface of leaves and on the stems and petioles, where the spermogonia also occur mixed with aecia.
- Aeciospores:are elliptical, yellowish brown, measure 14-22μ in diameter and have finely warty cell walls.
- Uredial stage: which is repeated several times during the season, produces spiny light brown uredospores.
- Urediniospores: are rounded, single, spinose, light brown, 21-30 x 18-26 microns.Pathogens produce several generations of urediniospores on peas.

- Telial stage: are not powdering, dark-brown, almost black.
- Teliospores: are unicellular, rounded, glabrous, positioned on colorless leg, 25-40 x 18-28 microns in size.It germinate to produce four celled basidia.
- Basidiospores: on which four sporidia are formed.
- Spores are spread by means of wind.

Disease Cycle

- Pathogen survives in soil and plant debris.
- Primary infection on the newly growing crop is considered to be caused by the germination of teleutospores present in soil, seed and collateral hosts in the month of January.
- Secondary infection by aeciospores or secondary aecia.

Epidemiology

- Frequent precipitations.
- Plentiful dews.
- Air temperature- 20-25 ^{0}C

Management

- Field sanitation.
- Growing of resistant varieties.
- Early sowing.
- Destruction of diseased plant debris.
- Crop rotation.
- Foliar spray of Carbendazim (0.1%) or Difoltan (0.3%) or Triadimefon 25% WP (0.1 %)given at 15 days intervals starting from first week of december.

Model Question Paper

Objective Type Questions

A. Objective Type Questions

a. Choose the correct answer from the following

(1) *Peronospora pisi* belongs to order.

a. Sphaeropsidales b. Peronosporales

c. Hyphomycetales d. Moniliales

(2) Downy mildew of pea is caused by :

a. *Peronospora pisi* b. *Rhizoctonia bataticola*

c. *Sphacelotheca sorghi* d. *Sphacelotheca cruenta*

(3) The *Peronospora pisi* causing downy mildew of peais

a. Facultative parasite b. Obligate parasite

c. Obligate saprophyte d. Fcultative saprophyte

(4) The pathogen *Peronospora pisi* survives in the soil and on old pea trash through

a. Oospore b. Perithecia

c. Apothecia d. Cleistothecia

(5) The fruiting body formed by *Erysiphe polygoni* in pea is

a. Pseudothecia b. Pycnidia

c. Apothecia d. Cleistothecia

(6) The secondary spreads of powdery mildew in pea is through

a. Soil borne conidia b. Air borne conidia

c. Seed borne conidia d. Vector

(7) The chemical wettable sulphur is found very effective for control of

a. Downy mildew b. Powdery mildew

c. Rust d. Late blight

(8) Which pathogen is an obligate parasite and ectophite, only haustoria enter into the cell

a. *Ascochyta rabiei*
b. *Rhizoctonia bataticola*
c. *Erysiphe polygoni*
d. *Sphacelotheca cruenta*

(9) *Uromyces pisi* is a:

a. Autocious rust
b. Heterocious rust
c. Dioecious rust
d. All

(10) Rust of pea is caused by

a. *Uromyces pisi*
b. *Rhizoctonia bataticola*
c. *Sphacelotheca sorghi*
d. *Sphacelotheca cruenta*

Answer

S. No	Answer	S. No	Answer
1	b. Peronosporales	7	b. Powdery mildew
2	a. *Peronospora pisi*	8	*c. Erysiphe polygoni*
3	b. Obligate parasite	9	a. Autocious rust
4	a. Oospore	10	*a. Uromyces pisi*
5	d. Cleistothecia		
6	b. Air borne conidia		

b. State whether the following statements are True or False

1) *Erysiphe polygoni* ia an obligate parasite.
2) Foliar spray with Metalaxyl + Mancozeb @ 0.2% manage downy mildew of pea.
3) White floury patches symptom appear on both side of the leaves of pea in powdery mildew disease.
4) The fungus *Fusarium oxysporum* f.sp. *pisi*produces three types of conidia like macroconidia, microconidia and chlamydospores.
5) Wilt disease of pea can be managed by soil application of Trichoderma.
6) The fungus *Peronospora pisi* has coenocytic mycelium.
7) The fungus *Peronospora pisi* is an obligate parasite.
8) The pathogen *Peronospora pisi* produce sex organs viz.,antheridia and oogonia in close proximity at the end of the pea season.

9) Primary infection of powdery mildew occurs in pea by asci and ascospores releases after breakdown of cleistothecia or conidia from collateral hosts.

10) Secondary spread of powdery mildew takes plae by wind borne conidia produced as a result of primary infection.

11) *Uromyces pisi* produces several generations of urediniospores on peas.

12) Telial stage of *Uromyces pisi* are dark-brown and almost black.

13) Primary infection of rust on the newly growing pea crop is caused by the germination of teleutospores present in soil, seed and collateral hosts in the month of January.

14) Secondary infection of rust on the newly growing pea crop is caused by by aeciospores or secondary aecia.

Answer

S. No	Answer	S. No	Answer
1	True	5	True
2	True	9	True
3	True	10	True
4	True	11	True
5	True	12	True
6	True	13	True
7	True	14	True
8	True		

B. Descriptive Questions

a. Long answer questions

1. Describe the causal organisms and mode of penetration of powdery mildew and downy mildew diseases of pea.
2. Describe in detail the occurrence, importance, symptoms, etiology, disease cycle and management of powdery mildew of pea.
3. Give the diagrammatic representation of disease cycle of the following diseases of pea

   i. Rust
   ii. Downy mildew
   iii. Anthracnose
   iv. Powdery mildew

3 Briefly describe the diagnostic symptoms and integrated management of any two of the following cotton diseases :

(i) Powdery mildew

(ii) Anthraconse

(iii) Wilt

5. Illustrate the powdery mildew of pea in the following headings

(i) Pathogen	(ii) Symptoms
(iii) Disease cycle	(iv) Disease management

2. Describe in detail the most distinguishing symptoms, causal organism, disease cycle and management of anthracnose of pea.

b. Short answer Questions

1. How will you differentiate Erysiphe from Sphaerotheca ?
2. Why the diseases caused by Erysiphe are called powdery mildew?
3. Which fungal pathogen causes powdery mildew of pea? Give its systematic position.
4. Give the characteristic of the oospore of *Peronospora pisi.*
5. What are cleistothecia.

c. Very short answer questions

1. Suggest some disease resistant varieties of pea against wilt disease.
2. Mention the name the resting structures by the following pathogens:

   (a) *Fusarium oxysporum* f.sp. *pisi*

   (b) *Colletotrichum pisi*

   (c) *Erysiphe pisi(d) Uromyces fabae*

3. List some important diseases and their causal organisms of pea ?
4. Distinguish between downy mildew and powdery mildew of pea.
5. Which type of disease is can be effectively managed by mixed cropping.

Fig. 1: Powdery mildew

Fig. 2: Downy mildew

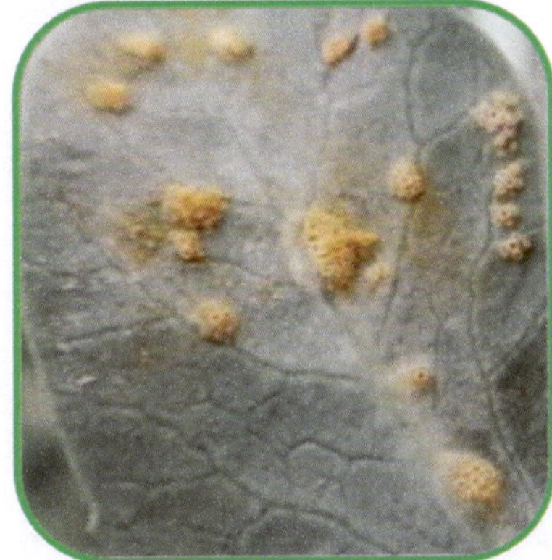

Fig. 3: Rust

Plate 10: Photograph showing symptoms of major diseases of pea

11

Black & Green Gram Crop Diseases & Management

S. No	Major Diseases	Causal Organism
1	Web blight	*Rhizoctonia solani*
2	Leaf spot	*Cercospora canescens*
3	Anthracnose	*Colletotrichum lindemuthianum*
4	Yellow mosaic	*Mungbean yellow mosaic virus*
5	Powdery mildew	*Erysiphe polygoni*
6	Rust	*Uromyces phaseoli*
7	Dry root rot	*Rhizoctonia bataticola*

Anthracnose

Diagnostic Symptoms

- Pathogen attacks all aerial parts of plants and at any stage of their growth.
- Symptoms are circular, black, sunken spots with dark center and bright red orange margins on leaves and pods.
- In severe infections, the affected parts wither off.
- Seedlings get blighted due to infection soon after seed germination.
- Infected pods have discolored seeds (Fig.3).

Etiology

Scientific Classification

Kingdom : Fungi

Division : Eumycota

Sub- division : Deuteromycotina

Class : Coelomycetes

Order : Melanconiales

Family : Melanconiaceae

Genus : *Colletotrichum*

Species : *lindemuthianum*

- Anamorphic stage:*Colletotrichum lindemuthianum*
- Sexual stage: *Glomerella lindemuthianum.*
- Pathogen mycelium is septate, hyaline and branched.
- Conidiophores are hyaline and short and bear cylindrical or oblong, hyaline, thin walled, single celled conidia with oil globules.
- Conidia are produced in acervuli, arise from the stroma beneath the epidermis and later rupture to become erumpent.
- A few dark coloured, septate setae are seen in the acervulus.
- Perfect stage of the fungus produces perithecia with limited number of asci.
- Asci contain typically 8 ascospores which are one or two celled with a central oil globule.

Disease Cycle

- Pathogens survive on seed and plant debris.
- Pathogen cause primary infection through seed.
- Secondary spread by air borne conidia produced on infected plant parts.
- Disease spreads in the fi eld through air-borne conidia.

Epidemiology

- Low temperature -15-20° C.
- High relative humidity - > 90 per cent.
- Cool rainy days.

Management

- Remove and destroy infected plant debris in soil.
- Seed treatment with Carbendazim @ 2 g/kg.
- Spray Carbendazim (0.1%) or Mancozeb (0.2%) soon after the appearance of disease and repeat after 15 days.

Cercospora Leaf Spot

Diagnostic Symptoms

- Infected leaves show small, brown, water soaked, circular spots surrounded with yellowish halo.
- On older plants the leaflet infection is mostly on older leaves and may cause serious defoliation. The most striking symptoms are on the green fruit.
- Small water-soaked spots first appear which later become raised and enlarge until they are one-eighth to one-fourth inch in diameter.
- Centres of these lesions become irregular, light brown and slightly sunken with a rough, scabby surface.
- Ripe fruits are not susceptible to the disease.
- Surface of the seed becomes contaminated with the pathogen inoculum, remaining on the seed surface for some time (Fig.2).

Etiology

Scientific Classification

Kingdom : Fungi

Division : Eumycota

Sub- division : Deuteromycotina

Class : Hyphomycetes

Order : Hyphomycetales (Moniliales)

Family : Dematiaceae

Genus : *Cercospora*

Species : *canescens*

- Anamorphi stage:*Cercospora canescens.*
- Produces clusters of dark brown septate conidiophores.
- Conidia are linear, hyaline, thin walled and 5-6 septate.

Disease Cycle

- Pathogen is soil borne on diseased plant debris and it survives only on the tissues which it colonizes as a parasite.
- Pathogen spreads about 3 m through the soil in one season, apparently along roots.
- Secondary spread is by air-borne conidia.

Epidemiology

- Moist weather and splattering rains.
- High humidity or persistent dew.

Management

- Field sanitation.
- Use disease free seed and resistant varieties.
- Adjustmet of sowing date.
- Intercrop the moong with tall growing cereals and millets.
- Maintain low crop population density and wide row planting.
- Foliar spray of Mancozeb (0.2%) or Carbendazim (0.1%) solution.
- Follow crop rotation.
- Destruciton of infected plant debris.

Web Blight

Diagnostic Symptoms

- Pathogen infects all above ground parts of the plants i.e. leaves, petioles, stem and pod.
- On the foliage, the symptoms start with the development of small, irregular, water soaked, greenish pale spots with damp appearance on any portion of leaf and stem.
- Spot turns dark brown, leaflets shrivel and dry up.
- Spot increases in size very rapidly covering greater parts of the leaf blade and stem, this is a very conspicuous and destructive phase of the disease.

- If infection occurs on collar region, reddish brown lesion in the cortex of hypocotyle at or below the soil level develops.
- Seedling is killed when the lesions on hypocotyle; girdles the stem after they coalesce or enlarge in size .
- Infected branches shrivel at the point of infection, leaves on it loose their normal green colour and wither.
- On young pods, the spots are light tan and irregular in shape but generally on mature pods, they are dark brown and sunken .
- Spots may coalesce to cover entire pod under favourable conditions.
- Coloured, spiderweb like mycelial growth and white microsclerotia develop in abundance on the affected plant parts.
- Sclerotia later turn brown to chestnutbrown in colour within 2-3 days of their development (Fig.1).

Etiology

Scientific Classification

Kingdom : Fungi

Division : Eumycota

Sub- division : Deuteromycotina

Class : Hyphomycetes

Order : Agonomycetales

Family : Agonomycetaceae

Genus : *Rhizoctoniza*

Species : *solani*

- Anamarphic stage; *Rhizoctonia solani*
- Teleomarphic stage : *Thanatephorus cucumeris*

Anamarphic stage; *Rhizoctonia solani*

- Colonies of *R. solani* are yellowish white to light yellowish brown and letter becoming pale brown to dark brown in colour.
- Sectoring of colony is frequent on agar medium.

- Rapidly growing mycelium; hyaline when young but later turning brown; 4.5 to 13.25 μ and branching near the distal septum of hyphal cell, often nearly at right angles in older hyphae.
- Cells of growing hyphae are always multinucleate and the septum is dolipore type.
- Sclerotia are superficial, variable in shape and size, dark brown to black, scab like and accompanied with dark short called, abundantly branched, stout mycelium without clamp connections.
- *Rhizoctonia solani* casual agent of web blight of mungbean belongs to anastomosis group 1, intraspecific group IB.

Teleomorphic stage : *Thanatephorus cucumeris*

- Perfect state *T. cucumeris* develops on healthy tissue adjacent to the lesion on black gram leaves as well as *in vitro*.
- Sexual stage is induced on the host under warm and humid conditions with heavy dew formation during the night.
- Fructifications are resupinate, creamy to greyish white, loosely attached to the substratum, composed of arachnoid, repent hyphae giving rise to thin hypochnoid or sub membranous fertile patches.
- Cluster of basidia are produced terminally in a discontinuous hymenium.
- Hyphae 4.5 to 17.0 μm wide without clamp connections, often anastomosed; basal hyphae widest with branching at wide angles while median and subhyaline hyphae are narrow, thin walled and hyaline.
- Basidia are mostly 9-25× 5-12.4 μm in size but some times short, wider, barrel shaped to sub cylindrical without a median construction.
- Sterigmata are straight, attenuate towards the apex usually bearing 4 basidium, rarely 2,3 or upto 7 in number.
- Basiospore are hyaline smooth thin walled, oblong to ellipsoid and dorsally fattened or broad ovoid and commonly widest at the distal end form secondary basidiospores .

Disease Cycle

- Pathogen is primarily soil borne and can survive for many years by producing sclerotia in soil and on plant tissue/seed coat.
- It germinate to produce vegetative threads (hyphae) of the fungus that can attack mungbean on to the leaves during heavy rains.

- Fungal hyphae will come in contact with the plants and become attached to its external surface.
- After attachment, the fungus continues to grow on the external surface of the plant and will cause disease by producing a specialized infection structure that the plant cell releases nutrients for continued fungal growth and development .
- As the fungus kill the plant cells, the hyphae continues to grow and colonize dead tissue, often forming sclerotia.
- Collateral weed hosts play an important role in initiation and early spread of the disease to the main host.
- Weed hosts are first to take the infection and facilitate the production of the basidiospores of the pathogen.
- There after the pathogen become air borne to cause leaf, stem or pod infection and produces fresh crops of basidiospores on the main host.

Epidemiology

- Higher aerial temperature -26-32 ^{0}C.
- Soil temperature- 30-33 ^{0}C .
- Relative humidity - 100% .

Management

- Field sanitation.
- Use disease free seed and resistant varieties.
- Seed treatment with Carbendazim or Thiophanate methyl @1g/kg seed.
- Foliar spray with Carbendazim or Thiophanate methyl @ 0.1%, if required repeat after 15 days.
- Shallow planting.
- Practice crop rotation.
- Intercrop the moong with tall growing cereals and millets.
- Follow wide row planting.
- Diseased material should be removed and destroyed.
- Weed hosts should be removed .

Yellow Mosaic

Diagnostic Symptoms

- Initially mild scattered yellow spots appear on young leaves.
- The next trifoliate leaves emerging from the growing apex show irregular yellow and green patches alternating with each other.
- Spots gradually increase in size and ultimately some leaves turn completely yellow.
- Infected leaves also show necrotic symptoms.
- Diseased plants are stunted, mature late and produce very few flowers and pods.
- Pods of infected plants are reduced in size and turn yellow in colour (Fig.4).

Etiology

- Disease is caused by *Mungbean yellow mosaic India virus* (MYMIV) in Northern and Central Region.It is caused by *Mungbean yellow mosaic virus* (MYMV) in western and southern regions.
- Begomovirus belonging to the family geminiviridae.
- Geminate virus particles, ssDNA, bipartite genome with two gemonic components DNA-A and DNA-B.

Disease Cycle

- Disease is transmitted by whitefly, *Bemisia tabaci*.
- Disease spreads by feeding of plants by viruliferous whiteflies.
- Weed hosts viz., *Croton sparsiflorus, Acalypha indica, Eclipta alba* and other legume hosts serve as reservoir for inoculum.
- Summer sown crops are highly susceptible.

Management

- Cultivate the crop during rabi season.
- Grow resistant black gram variety.

S. No	Crop	Resistant Varieties
1	Mungbean	ML131, ML267, PM-1, PM-2, PM-3, PM-4, PDM-54, PDM-139, BM-4, Mum-2, Narendra mung-1, TARM-1, Pusa Bold-1, Ganga-8
2	Urdbean	PDU-1, IPU-94-1, Pant U-19, UG-218, To 94-2, LBG-17, WBG-26, KU-300

- Seed treatment with Thiomethoxam-70WS or Imidacloprid-70WS @4g/kg.
- Soil application of Phorate, Disulfoton granules @1-2 kg a.i/ha in furrow during sowing.
- Spray Imidacloprid 17.8% SL @ 100 to 150 ml per acre or Thiamethoxam-25WG @ 100g per acre.
- Rogue out the diseased plants up to 40 days after sowing.
- Remove the weed hosts periodically.
- Increase the seed rate (25 kg/ha).

Model Question Paper

A. Objective Type Questions

a. Choose the correct answer from the following

1. Which one of the following diseases is caused by*Colletotrichum lindemuthianum* in green gram

    a. Anthracnose   b. Web blight

    c. Wilt   d. Downy mildew

2. Which one of the following pathogens causes Web blight of green gram.

    a. *Rhizoctonia solani*   b. *Cercospora arachidicola*

    c. *Sclerotinia sclerotiorum*   d. *Melampsora lini*

3. Which pathogen produces acervuli in green gram?

    a. *Peronospora parasitica*

    b. *Colletotrichum lindemuthianum*

    c. *Ustilago tritici*

    d. *Phytophthora infestans*

4. Asci of *Glomerella lindemuthianum* typically contains how many ascospores

   a. 5 b. 6

   c. 7 d. 8

5. The perfect stage of *Colletotrichum lindemuthianum* is

   a. *Glomerella lindemuthianum* b. *Peronospora parasitica*

   c. *Alternaria brassicae* d. *Sclerotinia sclerotiorum*

6. Anthracnose disease spreads in the mung bens fi eld mainly through

   a. Soil-borne conidia b. Air-borne conidia

   c. Water-borne conidia d. Seed-borne conidia

7. The coloured, spiderweb like mycelial growth and white microsclerotia develop in abundance on the affected green gram in which disease

   a. Anthracnose b. Web blight

   c. Wilt d. Downy mildew

8. The teleomorphic stage of *Rhizoctonia solani* is

   a. *Thanatephorus cucumeris* b. *Peronospora parasitica*

   c. *Alternaria brassicae* d. *Sclerotinia sclerotiorum*

9. *Rhizoctonia solani* casual agent of web blight of mungbean belongs to

   a. Anastomosis group 1, intraspecific group IB.

   b. Anastomosis group 2, intraspecific group IC.

   c. Anastomosis group 3, intraspecific group ID.

   d. Anastomosis group 4, intraspecific group IE.

10. Yellow mosaic disease in mungben in northern and central region is caused by

    a. *Mungbean yellow mosaic India virus (*MYMIV)

    b. *Mungbean yellow mosaic virus* (MYMV)

    c. *Yellow mosaic virus*

    d. *Tobacco mosaic virus*

11. Yellow mosaic disease in mungben in western and southern regions is caused by.

    a. *Mungbean yellow mosaic India virus*

    b. *Mungbean yellow mosaic virus*

    c. *Yellow mosaic virus*

    d. *Tobacco mosaic virus*

Answer

S. No	Answer	S. No	Answer
1	a. Anthracnose	6	b.Air-borne conidia
2	a. *Rhizoctonia solani*	7	b.Web blight
3	b. *Colletotrichum lindemuthianum*	8	a.*Thanatephorus cucumeris*
4	d.8	9	a. Anastomosis group 1, intraspecific group IB.
5	a. *Glomerella lindemuthianum*	10	a. *Mungbean yellow mosaic India virus* (MYMIV)
		11	*a.Mungbean yellow mosaic virus*

b. State whether the following statements are True or False

1. A few dark coloured, septate setae are seen in the acervulus of *Colletotrichum lindemuthianum.*
2. The perfect stage of the fungus *Colletotrichum lindemuthianum* produces perithecia with limited number of asci.
3. The pathogen *Glomerella lindemuthianum* cause primary infection in green gram through seed.
4. Cercospora leaf spot infected leaves show small, brown, water soaked, circular spots surrounded with yellowish halo.
5. The conidia of *Cercospora canescens* are linear, hyaline, thin walled and 5-6 septate.
6. The secondary spread of Cercospora leaf spot is through air-borne conidia.
7. The pathogen *Cercospora canescens* belongs to family dematicaeae.
8. Sectoring of colony of *Rhizoctonia solani* is frequently occur on agar medium.
9. Cells of growing hyphae of *Rhizoctonia solani* are always multinucleate and the septum is dolipore type.

10. The pathogen *Rhizoctonia solani* is primarily soil borne and can survive for many years by producing sclerotia in soil and on plant tissue.
11. The pathogen *Rhizoctonia solani* kill the plant cells, the hyphae continues to grow and colonize dead tissue, often forming sclerotia.
12. Yellow mosaic disease of black gram is transmitted in semi persistent manner by aphid *Aphis gossypii*.

Answer

S. No	Answer	S. No	Answer
1	True	7	True
2	True	8	True
3	True	9	True
4	True	10	True
5	True	11	True
6	True	12	True

B. Descriptive Questions

a. Long answer questions

1. Describe in detail the symptoms, etiology, disease cycle and management of anthracnose of black gram.
2. Describe the most distinguishing symptoms, causal organism, disease cycle and disease management of web blight disease of green gram.
3. Briefly describe the most distinguishing symptoms and disease management of any of the following plant disease :

   (i) Cercospora leaf spot of green gram

   (ii) Anthracnose of black gram
4. Describe web blight of green gram under the following headings.

   (i) Pathogen (ii) Symptoms

   (ii) Disease cycle (iv) Disease management
5. Describe in detail the symptoms, causal organism, mode of transmission and management of yellow mosaic of black gram.
6. Describe the most distinguishing symptoms, causal organism, disease cycle and disease management of cercospora leaf spot disease of green gram.

7. Describe the symptoms, causal organism, disease cycle and the management of any disease of green gram studied by you.

b. Short answer questions

1. Discuss the chemical management of Anthracnose of green gram.
2. Distinguish between leaf spot and leaf blight symptom.
3. What do you mean by web blight?
4. Write down the etiology of *Colletotrichum lindemuthianum.*
6. Which fungal pathogen cause cercospora leaf spot of green gram? Give its systematic position.
7. Elaborate disease cycle of cercospora leaf spot of black gram in pictorial form.
8. Describe in brief bout managment of cercosporaa leaf spot of black gram.

c. Very short answer questions

1. Name the pathogen causing the following plant diseases :

   (i) Anthracnose of green gram

   (ii) Cercospora leaf spot of black gram

   (iii) Web blight of green gram

   (iv) Powdery mildew of green gram

2. Name the disease caused by the following pathogens in mung and urd bean :

   (i) *Colletotrichum lindemuthianum*

   (ii) *Cercospora canescens*

   (iii) *Rhizoctonia solani*

   (iv) *Mungbean yellow mosaic virus*

3. Suggest cultural management of web blight of green gram.
4. Write down the etiology of *Rhizoctonia solani* causing web blight of green gram.

5. Write down the chemical management of cercospora leaf spot of black gram.

Fig. 1: Web blight

Fig. 2: Leaf spot

Fig. 3: Anthracnose

Fig. 4: Yellow mosaic

Plate 11: Photograph showing symptoms of major diseases of mungbean

12

Sugarcane Crop Diseases & Management

S. No	Major Diseases	Causal Organism
1	Red rot	*Colletotrichum falcatum*
2	Smut	*Ustilago scitaminea*
3	Grassy shoot	*Phytoplasma*
4	Ratoon stunting	*Clavibacterium xyli* sub sp. *xyli*
5	Pokkahboeng	*Fusarium moniliforme* and *Fusarium subglufinans*
6	Wilt	*Cephalosporium sacchari*
7	Leaf scald	*Xanthomonas albilineans*(Ashby)

Red Rot

Diagnostic Symptoms

- Practically all above ground parts are affected, but stalks are the main targets of stack.
- Spindle leaves (third or fourth)starts drying.
- Further stalks becomes discolored and hollow.
- Diagnostic symptom is th formation of red streaks, interrupted by white patches , inside the stalk.This can be seen by splitting the stalks longitudinally.
- White patches are characteristics of disease, not found in other stalk rots.
- After splitting opened the diseased stalk, a sour smell due to conversion of glucose to alcohol.
- Red longitudinal lesions are also seen on the mid ribs of the leaves on which black color fruiting body (Acervuli) are formed (Fig.1).

Etiology

Phylum	Deuteromycotina
Class	Hyphomycetes
Order	Melanconiales
Family	Melanconiaceae
Genus	*Colletotrichum*
Species	*falcaturm*

- Anamorphic stage:*Colletotrichum falcatum.*
- Perfect stage: *Physalospora tucumanensis.*
- Hyphae: profusely branched septate, hyphae containing oil droplets.
- Acervuli:black, minute velvettyacervuli with long, rigid bristle-like, septate setae.
- Conidiophores: a closely packed inside the acervulus, which are short, hyaline and single celled.
- Conidia: single celled, hyaline, falcate, granular and guttulate.
- Perithecia: dark brown to black with a papillate ostiole.
- Asci : clavate, unitunicate and eight-spored.
- Ascospore: hyaline, straight or slightly curved, ellipsoid or fusoid and unicellular which measure 18-22 μm x 7-8μm.

Disease Cycle

- Pathogen can survive chlaymydospores in crop debris or in soil.
- Pathogen produces acervuli and conidia in the soil that serve as primary inoculums and infect the sets. Diseased sets also give rise to infected plants.
- Secondary infection that spreads the disease is brought about by conidia formed on the mid rib lesions of the leaves.
- Dissemination of conidia is through wind , rain, heavy dew and irrigation waters.

Epidemiology

- Successive ratoon cropping.
- Monoculturing of sugarcane.
- Water logged conditions and injuries caused by insects.

- Low temperature.

Management

- Grow resistant and moderately resistant varieties CO 62198, CO 7704 , CO 8001, CO8201.
- Select the setts from the disease free fields or disease free areas.
- Setts can be treated with by moist hot air at 54°C for 2 hours and by aerated steam at 52 °C for 4 to 5 hours.
- Soak the setts in Carbendazim (0.1%) or Triademefon (0.05%) solution for 15 minutes before planting.
- Adopt crop rotation by including rice and green manure crops.
- Aviod ratooning of the diseased crop.

Smut

Diagnostic Symptoms

- Production of long whip-like structure from the gwoing point of of affected cane.
- Whip may be few millimeters to 20 mm in diameter.
- It is unbranched and made up of a fairly hard core of parenchyma and fibrovascular element in which the fungal hyphae ramify with the telioospore.
- Whip enclosed at first in a thin silvery sheath enclosing mass of black powdery spores.
- Initally thin canes with elongated internodes later become reduced in length.
- Profuse sporuting of lateral buds with narrow erect leaves especially in ratton crop (Fig.2).

Etiology

Phylum	Basidiomycotina
Class	Teliomyetes
Order	Ustilaginales
Family	Tilletiaceae
Genus	*Ustilago*
Species	*scitaminea*

- Pathogen: *Ustilago scitaminea*
- Pathogen is true culmicolous (stem infectiong) smut.
- Fungal hyphae are primarily intercellular and collect as a dense mass between the vascular bundles of host cell and produce tiny black spores.
- Teliospores are spherical light brown, and 10.5 to 17.04 m*m* in diameter.
- Teliospores germinate to produce 3-4 celled, hyaline promycelium and produce 4 sporidia (basidiospore) which are hyaline and oval shaped with pointed ends.
- Dikaryotic phase is established through fusion between sporidia, between cells of promycelium or between cells of hyphae developing from the germinating teliospores.
- Dikaryotic infection hypha invades the host tissue.
- When the whip is fully emerged the membrane ruptures releasing teliospores in the air.

Disease Cycle

- Pathogen survives as smut spores and dormant mycelium present in or on the infected setts and in the infected plant materials in the soil.
- Primary spread of the disease is through diseased seed-pieces (setts).
- Secondary spread in the field is mainly through the wind borne smut spores.
- Disease spread centrifugally in the field.

Epidemiology

- Continuous ratooning.
- Monoculture.
- Dry weather.

Management

- Plant healthy setts taken from disease free area.
- Grow resistant varieties like Co 7704, COC 85061 and COC 8201.
- Physical or chemical treatment of sets before planting.

- Physical method include hot water treatment, the sets are dipped in water at 55^{0}C to 60 ^{0}C for 10 minute, or exposure of sets to moist hot air at 54^{0}C for eight hours.
- Sett treatment with hot water at 50 ^{0}C plus Bayleton (0.1%) completely eliminate the smut from sets.
- Discourage ratooning of the diseased crops having more than 10 per cent infection.
- Remove and destory the smutted clump (collect the whips in a thick cloth bag/polythene bag and immerse in boiling water for 1 hr to kill the spores).
- Follow crop rotation with green manure crops or dry fallowing.

Grassy Shoot

Diagnostic Symptoms

- Characterized by proliferation of vegetative buds from the base of the cane giving rise to crowded bunch of tillers bearing narrow leaves.
- Leaves become pale yellow to completely chlorotic, thin and narrow.
- Plants appear bushy and 'grass-like' due to reduction in the length of internodes premature and continuous tillering.
- Cane formation rarely occurs in the affected clumps, if formed, thin with shorter internodes having aerial roots at the lower nodes (Fig.3).

Etiology

- Pathogen: *Phytoplasma*
- Two types of bodies are seen in ultrathin sections of phloem cells of infected plants.
- Spherical bodies of 300-400 nm diameter and filamentous bodies of 30-53 mm diameter in size.

Disease Cycle

- Pathogen is perpetuated through crop rattoning.
- Primary transmission isthrough infected seed material.
- Secondary spread is by aphid,grass hopper and dodder.

Management

- Select setts from diseased free area.
- Grow resistant varieties viz., Co 86249, CoG 93076 and Coc 22.
- Pre-treating the healthy setts with hot water at 52°C for 1 hour before planting.
- Complete cure of this disease was achieved by treating single-bud setts with 500 or 1000 ppm of Ledermycin or 250 ppm of Achromycin, Terrarriycin and Erythromycin.
- Rogue out infected plants in the secondary and commercial seed nursery.
- Spray Dimethoate @ 0.1 % to control insect vector.
- Avoid ratooning if GSD incidence is more than 15 % in the plant crop.

Ratoon Stunting

Diagnostic Symptoms

- It is called "ratoon" stunting because the disease is more severe on ratton crops than on 'plant cane' crops.
- It causes 5-15% loss in yield without the grower even noticing the disease because there no recognizable symptoms, except stunting, which too is not distinctive.
- Stunted growth, reduced tillering, thin stalks withshortened internodes and yellowish foliage.
- Show signs of wilting during summer.
- On spilliting open the cane longitudinally, two types of discoloration is seen in the pith,Orange-red vascular bundles in shades of yellow at the nodes are seen in the infected mature canes.
- Pink color is seen near the nodes in young canes (Fig.4).

Etiology

Kingdom	Bacteria
Division	Actinobacteria
Order	Actinomycetales
Family	Microbacteriaceae
Genus	*Clavibactor*
Species	*xyli*

- Pathogen*: Clavibacter xyli* sub sp. *xyli* (*Rickettsia like Organism - RLO)*
- Slow growing, gram positive, rod shaped.
- Pathogen is present in the xylem cells of infected plants so called as fastidious xylem colonizing bacterium.
- They are small, thin, rod shaped or coryneform, measuring 0.3-0.5x1-3 microns , with length of the filamentous eing 10 microns or more.
- Only sap transmitted and no insect vector is known.
- Disease spreads through cutting knives and seed setts from diseased canes.

Disease Cycle

- Disease is primarily transmitted through seed-pieces used for planting.
- Harvesting implements contaminated with infected juice also help spread the pathogen.
- Johnson grass, maize and Elephant grass act as carriers of the RSD bacterium and may serve as the source of noculums.

Epidemiology

- Optimum temperature -26-30°C.
- Moisture stress.

Management

- Select the setts from disease free fields .
- Treating the setts in hot water at 50°C for about 2 hours.
- Remove and burn the clumps showing the disease incidence.
- Remove inoculums of weak or stunted crops.

Pokkahboeng

Fusarium moniliforme and *Fusarium subglufinans*

Diagnostic Symptoms

- General symptoms of Pokkahboeng are mainly of three types
- Chlorotic Phase:

    - Chlorotic condition towards the base of the young leaves and occasionally on the other parts of the leaf blades.
    - Frequently, a pronounced wrinkling, twisting and shortening of the leaves accompanied the malformation or distortion of the young leaves.
    - Base of the affected leaves is seen often narrower than that of the normal leaves.
- Acute Phase or Top-Rot Phase:
    - Most serious stage of Pokkahboeng is a top rot phase. The young spindles are killed and the entire top dies.
    - Leaf infection sometimes continued to downward and penetrates in the stalk by way of a growing point.
    - In advanced stage of infection, the entire base of the spindle and even growing point showed a malformation of leaves, pronounced wrinkling, twisting and rotting of spindle leaves.
    - Red specks and stripes also developed.
- Knife-cut Phase :
    - Characterized by one or two or even more transverse cuts in the rind of the stalk /stem in such a uniform manner as if, the tissues are removed with a sharp knife.
- This is an exaggerated stage of a typical ladder lesion of a Pokkahboeng disease (Fig.5).

Disease Cycle

- Air-borne disease.
- Primarily transmitted through the air-currents.
- Secondary transmission is through the infected setts, irrigation water, splashed rains and soil.

Epidemiology

- Temperature- 20-30°C.
- Relative humidity ->70 to 80%.
- Cloudy weather, drizzling rains.

Management

- Plant healthy seed material.
- Use resistant varieties.
- Foliar spray with Carbendazim (0.1%) or Copper oxychloride(0.3%), Mancozeb (0.3%), if required repeat after 15 days.
- Canes showing 'top rot' or 'knife cut' should be rouged out .

Model Question Paper

A. Objective Type Questions

a. Choose the correct answer from the following

(1) Grassy shoot disease is caused by :

a. Phytoplasma b. Viroid

c. Prions d. Virus

(2) *Thanatephorus cucumaris* is the perfect state of :

a. *Collertrichum falcatum*

b. *Rhizoctonia solani*

c. *Rhizoctonia bataticola*

d. *Fusarium oxysporum* f. sp. *vasinfectum*

(3) *Ustilago scitaminea* causes :

a. Smut of sugarcane b. Wilt of cotton

c. Damping off of tobacco d. Wilt of sugarcane

(4) Red rot is a destructive disease of :

a. Cotton b. Tobacco

c. Sugarcane d. Potato

(5) The perfect state or teleomorph of *Colletotrichum falcatum* is

a. *Physalsospora tucumanensis* b. *Glomerella tucumanensis*

c. *Thanatephorus cucumeris* d. *Fusarium oxysporum*

(6) Ratoon stunting disease is caused by :

a. Phytoplasma b. Viroid

c. Bacteria d. Virus

(7) Red longitudinal lesions are also seen on the mid ribs of the leaves of sugarcane on which black color fruiting body are formed

a. Acervuli b. Pycnidia

c. Sporodochia d. Cleistothecia

(8) Diagnostic symptom is th formation of red streaks, interrupted by white patches , inside the stalk that can be seen by splitting the stalks longitudinally.

a. Red rot of sugarcane b. Smut of sugarcane

c. Grassy shoot of sugarcane d. Wilt of sugarcane

(9) Production of long whip-like structure from the growing point of affected sugarcane is a diagnostic symptom of.

a. Red rot b. Smut

c. Grassy shoot d. Wilt

(10) Which disese of sugarcane is characteriszed by proliferation of vegetative buds from the base of the cane giving rise to crowded bunch of tillers bearing narrow leaves

a. Red rot b. Smut

c. Grassy shoot d. Wilt

(11) Which disese of sugarcane is more severe on ratton crops than on plant cane crops.

a. Red rot b. Smut

c. Grassy shoot d. Ratoon stunting

S. No	Answer	S. No	Answer
1	a. Phytoplasma	7	a.Acervuli
2	b. *Rhizoctonia solani*	8	a.Red rot of sugarcane
3	a. Smut of sugarcane	9	a. Red rot
4	c. Sugarcane	10	c.Grassy shoot
5	a. *Physalsospora tucumanensis*	11	d.Ratoon stunting
6	a. Bacteria		

b. Fill in the blanks with suitable words

1. The perfect stage of *Colletotrichum falcatum* is ------------------------
2. After splitting opened the red rot diseased stalk, a sour smell due to conversion of glucose to ------------------------.
3. Red longitudinal lesions are also seen on the mid ribs of the leaves of sugarcane infected with red rot on which black color fruiting body------ ---------------- are formed
4. Setts can be treated with by moist hot air at ------------for 2 hours for management of red rot of sugarcane.
5. ------------------- is the resting spore of thesugarcane infecting fungus *Colletotrichum falcatum.*
6. Production of long whip-like structure from the gwoing point of of affected sugarcane is a symptom of ----------------disease .
7. *Ustilago scitaminea* belong to class---------------------.
8. When the whip is fully emerged in smut of sugarcane the membrane ruptures releasing -------------------in the air.
9. ------------------disease of sugarcane is characterized by proliferation of vegetative buds from the base of the cane giving rise to crowded bunch of tillers bearing narrow leaves.
10. Ratoon stunting disease is caused by -----------------------
11. --------------------------disease is more severe on ratton crops than on 'plant cane' crops.
12. The disese pokkahboeng is caused by-------------------------------.

b. Answer

S. No	Answer	S. No	Answer
1	*Physalospora tucumanensis.*	7	Teliomycetes
2	Alcohol	8	Teliospores
3	Acervuli	9	Grassy shoot
4	54°C	10	*Clavibacter xyli.*
5	Chlamydospore	11	Ratoon stunting
6	Smut	12	*Fusarium moniliforme*

c. True /Flse

1. Black, minute velvety acervuli with long, rigid bristle-like, septate setae are formed in sugarcane infected with red rot disease.
2. Dark brown to black perithecia with a papillate ostiole formed in sugarcane infected with pathogen *Colletotrichum falcatum.*
3. Secondary infection that spreads the red rot disease in sugarcane is brought about by conidia formed on the mid rib lesions of the leaves.
4. Monoculturing of sugarcane promote red rot disease in sugarcane.
5. Soak the setts in Triademefon (0.05%) solution for 15 minutes before planting manage red rot disease in sugarcane.
6. Discourage ratooning of the smut diseased sugarcane crops having more than 10 per cent infection.
7. *Clavibactor xyli* is a slow growing, gram positive, rod shaped bacteria.
8. *Clavibactor xyli* is present in the xylem cells of infected plants so called as fastidious xylem colonizing bacterium.
9. The pathogen *Clavibactor xyli* causing ratoon stunting is only sap transmitted and no insect vector is known.
10. Top-rot phase is the most serious stage of pokkahboeng disease.
11. Pokkahboeng of sugarcane is a air-borne disease.
12. Knife-cut phase is an exaggerated stage of a typical ladder lesion of a pokkahboeng disease.

c. Answer

S. No	Answer	S. No	Answer
1	True	7	True
2	True	8	True
3	True	9	True
4	True	10	True
5	True	11	True
6	True	12	True

B. Decriptive Question

a. Long answer questions

1. Describe in detail the diagnostic symptoms, etiology, disease cycle and management of red rot of sugarcane.

2. (a) Briefly describe the most distinguishing symptoms, causal organism and management of smut of sugarcane.

   (b) Briefly describe the most distinguishing symptoms, causal organism and management pokahboeng of sugarcane.

3. Describe in detail the most distinguishing symptoms causal organism, etiology, disease cycle and management of grassy shoot disease of sugarcane.

4. Describe in brief the most distinguishing symptoms, transmission and disease management of ratoon stunting.

5. Describe the disease pokkahboeng giving symptoms, pathogen disease cycle and management.

b. Short Answer Questions

1. Write the causal organism and major symptoms of the following disease:

   (i) Smut of sugarcane

   (ii) Grassy shoot of sugarcane

2. Describe in brief about disease management of red rot of sugarcane.

3. Explain the symptom and management of smut of sugarcane.

4. Explain disease cycle, epidemiological condition and management of pokkahboeng disese of sugarcane.

5. Give digramatic representation of disease cycle of smut of sugarcane and their management.

6. Explain the symptom and management of grassy shoot of sugarcane.

c. Very Short Answer Questions

1. Name the pathogen causing the following plant disease:

   (i) Red rot of sugarcane (ii) Pokkahboeng

   (iii) Wilt of sugarcane (iv) Ratoon stunting of sugarcane

2 Name the disease caused by the following pathogens :

   (i) *Cephalosporium sacchari*

   (ii) *Xanthomonas albilineans*

   (iii) *Colletotrichum falcatum*

   (iv) *Clavibacterxyli*sub sp. *xyli*

3. Write down the Integrated management schedule of whip smut of sugarcane.
4. What do you mean by grassy shoot?
5. What do you mean by Pokkahboeng?

Fig. 1: Red rot

Fig. 2: Smut

Fig. 3: Grassy shoot

Fig. 4: Ratton stuting

Fig. 5: Pokkahboeng

Plate-12: Photograph showing symptoms of major diseases of Sugarcane

13

Mustard Crop Diseases & Management

Major Diseases	Causal Organism
Alternaria blight	*Alternaria brassicae*, *A. brassicicola*, and *A.raphani*
White rust	*Albugo candida*
Downy mildew	*Peronospora parasitica*
Sclerotinia stem rot	*Sclerotinia sclerotiorum*
Powdery mildew	*Erysiphe cruciferarum*
Club root	*Plasmodiophora brassicae*
Bacterial blight/ Black rot	*Xanthomonas campestris pv. Campestris*

Alternaria Blight

Diagnostic Symptoms

- Symptom first appear on the lower leaves as small circular brown necrotic spots which slowly increase in size.
- Many concentric spots coalesce to cover large patches showing blightening and defoliation in severe cases.
- Circular to linear, dark brown lesions also develop on stems and pods, which are elongated at later stage.
- Infected pods produce small, discolored and shriveled seeds(Fig.2).

Etiology

Scientific Classification

Kingdom : Fungi

Phylum : Ascomycota

Class : Dothideomycetes

Subclass : Pleosporomycetidae

Order : Pleosporales

Family : Pleosporaceae

Genus : *Alternaria*

Species : *brassicae/brassicicola*

Alternaria brassicae

- Mycelium: Immersed, hyphae branched, septate, hyaline, smooth,4-8μm thick.
- Conidiophores: arising in groups of 2-10 or more from the hyphae which are straight or curved.
- Conidia: light olive coloured with transverse and longitudinal septa. These are around 3-5 septate and conidia are borne in chain.

Alternaira brassicicola

- Mycelium: immersed, hyphae branched, septate, hyaline, olivaceous brown, smooth, upto 70 μm long and 4-8μm thick.
- Conidiophores: arising in groups of 2-12, emerging through stomata, usually simple, erect, straight or curved, smooth upto 70μm long, 5-8 μm thick.
- Conidia: mostly in chains of upto 20 or more, sometimes branched, straight , nearly cylindrical, usually slightly tapering towards the apex, or obclavate, the basal cell rounded, the beak usually almost nonexistent, are light olive coloured with1-11 mostly less than 6 transverse septa.

Disease Cycle

- Pathogens survive between mustard and canola crops in infested crop debris and on seed.
- These pathogens are disseminated as conidia within and among fields by splashing water and wind.
- Pathogens infect their hosts by direct penetration or through wounds andnatural openings.

Epidemiology

- Warm and humid weather.
- Relative humidity-> 70 percent.
- Optimum temperature-12-20 ^{0}C.
- Intermittent rain.

Management

- Plant high quality seed free from the Alternaria black spot pathogens.
- Practice a three year or longer rotation to non-hosts such as small grains.
- Promptly incorporate crop residues after harvest to hasten their breakdown.
- Eliminate volunteer canola, mustard and cruciferous weeds.
- Early sowing in first fortnight of october.
- Spray the crop with Mancozeb (0.2%) or Iprodione (0.2%).

White Rust

Diagnostic Symptoms

- Initially chlorotic (yellowed) lesions and sometimes galls appeared on the upper leaf surface.
- There are corresponding white blister-like dispersal pustules of sporangia on the underside of the leaf.
- Branches and flower parts deformed and pathogen stimulates hypertrophy and hyperplasia.
- Flower and young fruits may also malformed and malformed pod lack seed development.
- Disease is sometimes called the cancer of rapeseed as one phase of this disease shows grotesque malformation of the young shoots and inflorescence (Fig.1).

Etiology

Scientific Classification

Phylum : Heterokontophyta

Class : Oomycota

Order : Peronosporales

Family : Albuginaceae

Genus : *Albugo*

Species : *candida*

- Pathogen :*Albugo candida.*
- Pathogen is obligate parasite and it needs a living host to grow and reproduce.
- Mycelium: Intercellular producing small globose, knob shaped haustoria in the host cells.
- Sporangiophores: hyaline, clavate, free from each other laterally and are very thick walled especially towards base.
- Sporangia: hyaline, nearly oval to spherical with uniform thin walled and 14-16 x16-20µm in size.It may germinate by means of germ tube or through zoospores.
- Zoospores: Biflagellate, lose their flagella, come to rest, and then germinate by germtube which penetrates the epidermis of the host directly or through the stomata.
- Oogonium: globose, terminal or intercalary and multinucleate.
- Antheridium :clavate, paragynous and contains 6-12 nuclei.
- Oospores : globose, chocolate brown,30-35µm in diameter. It germinate by formation of vesicles in which zoospores are formed or the oospores may germinate by production of one or two simple or branched germ tube by release of zoospores from sessile vesicles.
- Germination of sessile vesicles is most common.

Disease Cycle

- Pathogen reproduce by producing both sexual spores (called oospores) and asexual spores (called sporangia) in a many-stage (polycyclic) disease cycle.
- Thick-walled oospores are the main overwintering structures, but the mycelium can also survive in conditions where all the plant material is not destroyed during the winter.
- In the spring, the oospores germinate and produce sporangia (s. sporangium) on short stalks called sporangiophores that become so tightly packed within the leaf that they rupture the epidermis and are consequently spread by the wind.
- Liberated sporangia can either germinate directly with a germ tube or begin to produce biflagellate motile zoospores.

- Zoospores then swim in a film of water to a suitable site and each one produces a germ tube – like that of the sporangium – that penetrates the stoma.
- When the oomycete has successfully invaded the host plant, it grows and continues to reproduce.

Epidemiology

- Cool and moist condition.
- Optimal temperature – 13 – 25 °C.
- Light rain or irrigation.
- Leaf surfaces need to remain wet for at least 2 to 3 hours .

Management

- Sanitation.
- Use of resistant cultivars.
- Early sowing preferably in the first fortnight of October.
- Use the seed from stag-head free plants to avoid carry over of oospore through seeds.
- Treat the seed with Apron 35 SD @ 6gm/kg seed.
- Foliar spray the crop with Mancozeb (0.2%) at the onset of the disease, repeat the spray after 15 days interval.
- Destroy the diseased plant debris.
- Crop rotation.

Downy Mildew

Diagnostic Symptoms

- Intially grayish white irregular necrotic patches develop on the lower surface of leaves.
- Later under favorable conditions, brownish white fungal growth may also be seen on the spots.
- Most conspicuous and pronounced symptom is the infection of inflorescence causing hypertrophy of the peduncle of inflorescence and develop stag head structure (Fig.3).

Etiology

Scientific Classification

Kingdom : Fungi

Division : Eumycota

Class : Oomycetes

Order : Peronosporales

Family : Peronosporaceae

Genus : *Peronospora*

Species : *parasitica*

- Pathogen*: Peronospora parasitica.*
- Obligate parasite.
- Mycelium: strictly intercellular, hyaline with large, finger shaped haustoria.
- Sporangiophores: 100-300µm long and unbranched for a major portion of their length.Dichotomous branching occurs, 6-8 times, at the tip. They are at acute angle with each other. A single sporangium is borne at the tip of each ultimate branch.
- Sporangia: dispersed by rain splash or wind and may germinate by means of germ tube
- Oogonia: pale yellow, irregularly round, and swollen into crest like folds.
- Antheridia: tendril like and are produced on separate hyphae.
- A fertilization tube grows from the antheridium through the receptive papilla towards a central body in the ooplasm to discharge a single male nucleus.The two nuclei fuse and initiate the uninucleate oospore.
- Oospores: globose, measuring 26-43µm in diameter and germinate by germ tube .

Disease Cycle

- Pathogen survives between crops in the soil as oospores.
- Infection occurs when soilborne resting structures called oospores germinate and produce sporangia under moist, cool conditions.

- Sporangia are produced on the underside of leaves in the evening and are released during the day as leaves dry.
- Windblown sporangia land on leaves and directly penetrate leaves and flowers.

Epidemiology

- Cool, moist conditions.
- Optimum temperature-15-20°C.
- Relative humidity-70 percent.

Management

- Plant resistant or tolerant varieties.
- Seed treatment with Apron SD-70 (2g/kg seed).
- Sow the crop on normal planting date.
- Destroy diseased plant parts.
- Foliar spray with Mancozeb (0.2%) orRidomil MZ (02%), If needed repeat after 15 days of first spray.

Sclerotinia Stem Rot

Diagnostic Symptoms

- Initial symptoms appear as the plant canopy closes over rows during flowering and pod development.
- Lodged stems in contact with the soil develop watery lesions, with snowy white mycelium and black, irregularly shaped sclerotia become apparent as disease progresses.
- When the stem is completely girdled by such lesions, the plants wilt and die.
- When the crop is at seed stage, the plants tend to lodge, touching the siliquae to the soil level.
- Such plant show rotting of the siliquae with fungal growth along with the sclerotial, bodies just above the soil level(Fig.4).

Etiology

Scientific Classification

Kingdom : Fungi

Division : Ascomycota

Class : Leotiomycetes

Order : Helotiales

Family : Sclerotiniaceae

Genus : *Sclerotinia*

Species : *sclerotiorum*

- Pathogen *:Sclerotinia sclerotiorum.*
- Pathogen is characterized by the formation of hard blackish sclerotinia which germinate and produce cup shaped brown color apothecia.
- Sclerotia borne superficially usually on dense white mycelium, globose to cylindrical, but quite variable in shape, 2-15x2-15μm with black outer rind and white inner contex.
- Apothecia arising one to several form a sclerotium ochraceous often darker at the base of the stipe, receptacle 2-8 mm broad applanate to slightly concave, often with control depression frequently with an undulate margin, tapering to form a stipe 3-10 mm long, 1-2 mm wide.

Disease Cycle

- Pathogen survives between crops in the soil as sclerotia.
- Infection occurs when soilborne sclerotia (dormant resting structures) germinate when the soil surface is continuously wet for at least two weeks.
- Germinating sclerotia form small, tan, cup-shaped structures called apothecia, which release millions of airborne ascospores that colonize dead plant parts (such as senescent flowers) and infect host tissues.
- Sclerotia can survive in soil for up 8 to 10 years.

Epidemiology

- Cool wet weather.
- Optimum temperature for apothecia formation-15^{0}C

- Continuous leaf wetness is required for infection by ascospores-App. 48-72h.
- Dense plant stands.

Management

- Plant high quality seed free from sclerotia.
- Practice a four-year or longer rotation to non-hosts such as corn or small grains.
- Avoid excessive irrigation and fertilization.

Model Question Paper

A. Objective Type Questions

a. Choose the correct answer from the following

(1) Which one of the following diseases is caused by *Peronospora parasitica.*

a. Early leaf spot of groundnut b. Late leaf spot of groundnut
c. Wilt of linseed d. Downy mildew of mustard

(2) Which one of the following pathogens causes white rust of mustard.

a. *Albugo candica* b. *Cercospora arachidicola*
c. *Sclerotinia sclerotiorum* d. *Melampsora lini*

(3) Which pathogen reproduce by sexual spores called oospores?

a. *Peronospora parasitica* b. *Mycospharella berkeleyii*
c. *Gibberella indica* d. None of these

(4) Which one of the following is an obligate parasite.

a. *Albugo candida* b. *Melampsora lini*
c. *Fusarium oxysporum* f. sp. *lini* d. *Mycosphaerella arachidis*

(5) Which one of the following has light olive coloured conidia with transverse and longitudinal septa?

a. *Albugo candida* b. *Peronospora parasitica*
c. *Alternaria brassicae* d. *Sclerotinia sclerotiorum*

(6) Sclerotinia stem rot disease is caused by

a. Phytoplasma
b. Viroid
c Fungi
d. Virus

(7) Light olive coloured conidia with transverse and longitudinal septa

a. *Colletotrichum falcatum*
b. *Alternaria brassicae*
c. *Rhizoctonia bataticola*
d. *Fusarium oxysporum* f. sp. *vasinfectum*

(8) White blister-like dispersal pustules of sporangia on the underside of the leaf of mustard is characteristic symptom of disesase

a. White rust
b. Alternaria blight
c. Damping off
d. Downy mildew

(9) Which disease is sometimes called the cancer of rapeseed

a. White rust
b. Alternaria blight
c. Damping off
d. Downy mildew

(10) Which pathogen is an obligate parasite and it needs a living host to grow and reproduce.

a. *Albugo candida*
b. *Glomerella tucumanensis*
c. *Thanatephorus cucumeris*
d. *Fusarium oxysporum*

(11) Thick-walled oospores are the main overwintering structures found in

a. *Albugo candida*
b. *Glomerella tucumanensis*
c. *Thanatephorus cucumeris*
d. *Fusarium oxysporum*

(12) *Peronospora parasitica* survives between crops in the soil a

a. Acervuli
b. Pycnidia
c. Oospores
d. Cleistothecia

(13) Most conspicuous and pronounced symptom of which disease of mustard is the infection of inflorescence causing hypertrophy of the peduncle of inflorescence and develop stag head structure

a. White rust
b. Alternaria blight
c. Damping off
d. Downy mildew

(14) Foliar spray with Ridomil MZ @ 0.2% manage

a. Rust
b. Alternaria blight
c. Powdery mildew
d. Downy mildew

(15) Mustard lodged stems in contact with the soil develop watery lesions, with snowy white mycelium and black, irregularly shaped sclerotia become apparent as which disease progresses.

a. White rust
b. Alternaria blight
c. Sclerotinia stem rot
d. Downy mildew

Answer

S. No	Answer	S. No	Answer
1	d. Downy mildew of mustard	9	a. White rust
2	a. *Albugo candica*	10	a. *Albugo candida*
3	a. *Peronospora parasitica*	11	a. *Albugo candida*
4	a. *Albugo candida*	12	c. Oospores
5	c. *Alternaria brassicae*	13	d. Downy mildew
6	c. Fungi	14	d. Downy mildew
7	b. *Alternaria brassicae*	15	c. Sclerotinia stem rot
8	a. White rust		

b. State whether the following statements are True or False

1. The pathogen *Sclerotinia sclerotiorum* survives between crops in the soil as sclerotia.
2. *Sclerotinia sclerotiorum* infection occurs in mustard when soilborne sclerotia germinate when the soil surface is continuously wet for at least two weeks.
3. Presence of concentric ring and yellow halo is characteristic symptom of Alternaria blight of mustard.
4. *Albugo candida* stimulates hypertrophy and hyperplasia in the branches of mustard.
5. Germination of sessile vesicles is most common phenomena in *Albugo candida.*
6. The pathogen *Albugo candida* reproduce by producing both sexual and asexual spores in a many-stage of their disease cycle.

7. *Peronospora parasitica* infection occursin mustard when soilborne resting structures called oospores germinate and produce sporangia under moist, cool conditions.
8. The pathogen *Sclerotinia sclerotiorum* is characterized by the formation of hard blackish sclerotinia which germinate and produce cup shaped brown color apothecia.
9. Alternaria blight disease are disseminated within and among mustrd fields by splashing water and wind.
10. Germinating sclerotia of *Sclerotinia sclerotiorum* form small, tan, cup-shaped structures called apothecia.
11. Sclerotia of *Sclerotinia sclerotiorum* can survive in soil for up 8 to 10 years.

Answer

S. No	Answer	S. No	Answer
1	True	7	True
2	True	8	True
3	True	9	True
4	True	10	True
5	True	11	True
6	True		

B. Descriptive Questions

a. Long answer questions

1. Describe in detail the symptoms, etiology, disease cycle and management of white rust of mustard.
2. Describe the most distinguishing symptoms, causal organism, disease cycle and management of downy mildew disease of mustard.
3. Briefly describe the most distinguishing symptoms and management of any of the following plant disease :

   (i) Alternaria blight of mustard

   (ii) Sclerotinia stem rot of mustard
4. Describe white rust of mustard under the following headings.

(i) Pathogen (ii) Symptoms

(ii) Disease cycle (iv) Disease management

5. Describe in detail the symptoms, causal organism, disease cycle and management of powdery mildew of mustard.
6. Describe the symptoms, causal organism, disease cycle and the management of any disease of mustard studied by you.

b. Short answer questions

1. Give the chemical management of white blisters of white rust of mustard.
2. Distinguish between powdery and downy mildew of mustard.
3. Describe the hypertrophied symptoms of white blisters of mustard.
4. What is etiology? Write down the etiology of *Albugo candida* causing white blister in mustard.
5. Discuss about disease cycle and epidemiology of Sclerotinia stem rot diseases of mustard.

c. Very short answer Questions

1. Name the pathogen causing the following plant diseases :

   (i) Downy mildew of mustard (ii) Powdery mildew of mustard

   (iii) White blisters of mustard (iv) Alternaria blight of mustard
2. Name the disease caused by the following pathogens :

   (i) *Albugo candida* (ii) *Peronospora parasitica*

   (iii) *Alternaria brassicae* (iv) *Sclerotinia sclerotiorum*
3. Suggest cultural management of Sclerotinia stem rot disease of mustard.
4. Write down the etiology of *Peronospora parasitica* causing downy mildew of mustard.
5. Write down the chemical management of Alternaria blight of mustard.

Fig. 1: White rust

Fig. 2: Downey mildew

Fig. 3: Alternaria leaf spot

Fig. 4: Sclerotinia stem rot

Plate 13: Photograph showing symptoms of major diseases of mustard

14

Sunflower Crop Diseases & Management

Major Diseases	Causal Organism
Sclerotinia stem rot	*Sclerotinia sclerotiorum* (Lib.) de Bary
Alternaria blight	*Alternaria helianthi* (Hansf.) Tubaki& Nishi
Rust	*Puccinia helianthi* Schwein
Downy mildew	*Plasmopara halstedii* Farl. Berl. & De Toni
Necrosis	Tobacco streak virus
Head rot	*Rhizopus arrhizus* Fischer
Root rot or charcoal rot	*Macrophomina phaseolina* (Tassi) Goid

Sclerotinia Stem Rot

Diagnostic Symptoms

- Early symptoms of the disease are noticed forty days after sowing.
- Stem is usually infected at or near the soil line.
- Brownish lesion develops at the base of the stem and eventually girdles the plant. A white fan-like inoculum growth forms over the infected tissues and often radiates over the soil surface.
- Silky appearance of plants can be spotted from a distance and a row effect can be observed in heavily infected soil. Later the entire plant withers and dies.
- The lesion grows up the stem, destroying the cortical tissue and leaving the fibrous vascular strands as the tissues dry out.
- White cottony mycelium and mustard seed type sclerotial bodies are conspicuous on the affected stem near soil level. This also causes collar rot in the seedling stage (Fig.2).

Etiology

Scientific Classification

Kingdom : Fungi

Division : Ascomycota

Class : Leotiomycetes

Order : Helotiales

Family : Sclerotiniaceae

Genus : *Sclerotinia*

Species : *sclerotiorum*

- *Pathogen: Sclerotinia sclerotiorum.*
- Pathogen form hard blackish sclerotinia which germinate and produce cup shaped brown color apothecia.
- Sclerotia borne superficially usually on dense white mycelium, globose to cylindrical, but quite variable in shape, 2-15x2-15µm with black outer rind and white inner contex.
- Apothecia arising one to several form a sclerotium ochraceous often darker at the base of the stipe, receptacle 2-8 mm broad applanate to slightly concave, often with control depression frequently with an undulate margin, tapering to form a stipe 3-10 mm long, 1-2 mm wide.

Disease Cycle

- Pathogen survives between crops in the soil as sclerotia.
- Primary infection occurs when soilborne sclerotia germinate.
- Germinating sclerotia form small, tan, cup-shaped structures called apothecia, which release millions of airborne ascospores (secondary infection) that colonize dead plant parts and infect host tissues.

Epidrmiology

- Cool wet weather.
- Optimum temperature-15^{0}C.
- Leaf wetness –Approx. 48-72h.
- Dense plant stands.

Management

- Deep ploughing during summer.
- Crop rotation of 3-4 years.
- Timely sowing should be done.
- For resistant/ tolerant varieties.
- Soil amendment with FYM @ 5 tonnes/acre.
- Avoid moisture stress during high summer and water logging conditions.
- Seed treatment with Trichoderma @ 6g/kg seed.
- Addition of Trichoderma in soil at 10g/kg and soil amendments like castor cake, neem cake, oat straw reduces disease incidence.

Alternaria Blight

Diagnostic Symptom

- Circular to oval, dark brown to black spots appear on the leaves.The spots enlarge into round to irregular spots with concentric rings and coalesce causing blighting and withering of leaves.
- These are surrounded by chlorotic zone with grey white necrotic center.
- Spot appear first on the lower leaves and as the plant grows the spots subsequently appears on middle and upper leaves.
- Stem lesions begin as dark flecks which enlarge to form long, narrow lesions which may also coalesce to a larger blackened area resulting in stem breakage.
- In severe cases, the lesions appear on petioles, ray florets and head as brown round spots about 1cm diameter with a slight depression in the centre.
- Sometimes rotting of flower heads also occur (Fig.1).

Etiology

Phylum	Duteromycotina
Class	Hyphomycetes
Order	Hyphomycelates
Family	Dematicaeae
Genus	*Alternaria*
Species	*helianthi*

- Pathogen: *Alternaria helianthi.*
- Facultative parasite.
- Pathogen produces cylindrical conidiophores and conidia.
- Conidiophores are pale grey-yellow coloured, straight or curved, geniculate, simple or branched, septate and bear single conidium.
- Conidia are cylindrical to long ellipsoid, straight or slightly curved, 1 to 2 septate with longitudinal septa and pale grey-yellow to pale brown and size are 25-80 (120) x 8-11 microns.

Disease Cycle

- Pathogen overwinters as mycelium on infected plant residues and in dry conditions survives for 20 weeks in so.
- Primary source of inoculums are spores present in the soil and infected seed.
- Secondary spread is mainly through windblown conidia.

Epidemiology

- Air temperatures – 25 to 30 ^{0}C.
- Rainy weather.
- Late sown crops are highly susceptible.

Management

- Field sanitation.
- Deep summer ploughing.
- Use of resistant or tolerant variety like B.S.H.1 .
- Seed treatment with Thiram or Carbendazim at 2 g/kg.
- Spray Mancozeb (0.2%) or Copper oxychloride,(0.3%) Difenoconazole (0.1%) or Chlorothaloni l(0.2%).
- Proper spacing.
- Application of well rotten manures.
- Practicing crop rotation.
- Planting in mid-september.
- Remove and destroy the diseased plants.

Model Question Paper

A. Objective Type Questions

a. Choose the correct answer from the following

(1) *Sclerotinia sclerotiorum* belongs to order.

a. Helotiales b. Peronosporales

c. Hyphomycetales d. Moniliales

(2) The pathogen *Sclerotinia sclerotiorum* form hard blackish sclerotinia which germinate and produce cup shaped brown color

a. Apothecia *b.* Perithecia

c. Cleistothecia d. None

(3) Circular to oval, dark brown to black spots appear on the leaves which enlarge into round to irregular spots with concentric rings and coalesce causing blighting and withering of leaves are the diagnostic symptom of disease

a. Head rot b. Downy mildew

c. Necrosis d. Alternaria blight

(4) Which one of the following diseases is caused by *Rhizopus arrhizus* in sunflower *?*

a. Head rot b. Downy mildew

c. Necrosis d. Powdery mildew

(5) Which one of the following pathogens cause Alternaria blight of sunflower?

a. *Puccinia helianthi* b. *Alternaria helianthi*

c. *Cercosporidium personatum* d *Melampsora lini*

(6) Which pathogen form hard blackish sclerotinia which germinate and produce cup shaped brown color apothecia.

a. *Puccinia helianthi* b. *Alternaria helianthi*

c. *Sclerotinia sclerotiorum* d. *Melampsora lini*

(7) Which one of the pathogen causes sunflower necrosis?

a. Tobacco streak virus
b. *Puccinia helianthi*
c. *Sclerotinia sclerotiorum*
d. *Mycosphaerella spp.*

(8) Fungicides used for the management of Alternaria blight of sunflower.

a. Mancozeb
b. Copper oxychloride
c. Difenoconazole
d. All of the above

(9) Stem is usually infected at or near the soil line. Brownish lesion develops at the base of the stem and eventually girdles the plant.

a. Sclerotinia stem rot
b. Downy mildew
c. Necrosis
d. Powdery mildew

(10) The pathogen *Puccinia helianthi* belongs to the family

a. Dematiaceae
b. Albuginaceae
c. Pythiaceae
d. Erysiphacee

Answer

S. No	Answer	S. No	Answer
1	a. Helotiales	6	c. *Sclerotinia sclerotiorum*
2	a. Apothecia	7	a. Tobacco streak virus
3	d. Alternaria blight	8	d. All of the above
4	a. Head rot	9	a. Sclerotinia stem rot
5	b. *Alternaria helianthi*	10	a. Dematiaceae

b. State whether the following statements are True or False

1. Silky appearance can be spotted from a distancein a sclerotinia wilt and stem rot infected sunflower plants and a row effect can be observed in heavily infected soil.
2. In Alternaria blight of sunflower, spots appear first on the lower leaves and as the plant grows the spots subsequently appears on middle and upper leaves.
3. White cottony mycelium and mustard seed type sclerotial bodies are conspicuous on the sclerotinia wilt and stem rot affected sunflower stem near soil level.
4. *Alternaria helianthi* is a facultative parasite.

5. Sclerotia borne superficially usually on dense white mycelium of *Sclerotinia sclerotiorum*, globose to cylindrical, but quite variable in shape with black outer rind and white inner contex.
6. The pathogen *Sclerotinia sclerotiorum* survives between crops in the soil as sclerotia.
7. Germinating sclerotia of *Sclerotinia sclerotiorum* form small, tan, cup-shaped structures called apothecia, which release millions of airborne ascospores that colonize dead plant parts and infect host tissues.
8. The primary source of inoculums of *Puccinia helianthi* are spores present in the soil and infected sunflower seed.
9. The secondary spread of Alternaria blight in sunflower is mainly through windblown conidia.

b. Answer

S. No	Answer	S. No	Answer
1	True	6	True
2	True	7	True
3	True	8	True
4	True	9	True
5	True		

B. Descriptive Questions

a. Long answer questions

1. Describe in detail the symptoms, etiology, disease cycle and management of Alternaria blight sunflower.
2. Briefly describe the most distinguishing symptoms, causal organism, disease cycle and management of Sclerotinia stem rot disease of sunflower.
3. Briefly describe the most distinguishing symptoms, etiology and management of any of the following plant disease :

   (i) Alternaria blight of sunflower

   (ii) Sclerotinia stem rot of sunflower.
4. Describe White blisters and Alternaria blight of sunflower under the following headings.

(i) Etiology (ii) Symptoms

(ii) Favorable symptoms (iv) Disease management

5. Describe in detail the symptoms, causal organism, disease cycle and management of rust of sunflower.

b. Short answer questions

1. Describe chemical management of Alternaria blight of sunflower.
2. Write etiology and disease cycle of pathogen causing Alternaria blight of sunflower.
3. Describe the diagnostic symptoms and management of Sclerotinia stem rot of sunflower.
4. Mention diagnostic symptoms, causal organism and etiology of Alternaria blight of sunflower.
5. Describe the disease cycle and epidemiology of Sclerotinia stem rot of sunflower.

c. Very short answer questions

1. Name the pathogen causing the following plant diseases :

   (i) Rust of sunflower

   (ii) Alternaria blight of sunflower

   (iii) Sclerotinia stem rot of sunflower

   (iv) Downy mildew of sunflower
2. Give the name of different types of spore stages found in *Puccinia helianthi*.
3. Name the disease caused by the following pathogens :

   (i) *Macrophomina phaseolina* (ii) *Alternaria helianthi*

   (iii) *Sclerotinia sclerotiorum* (iv) *Plasmopara halstedii*
4. Suggest at least four resistant varieties of sunflower against Alternaria blight disease of sunflower.
5. Write etiology of pathogen causing Sclerotinia stem rot of sunflower.

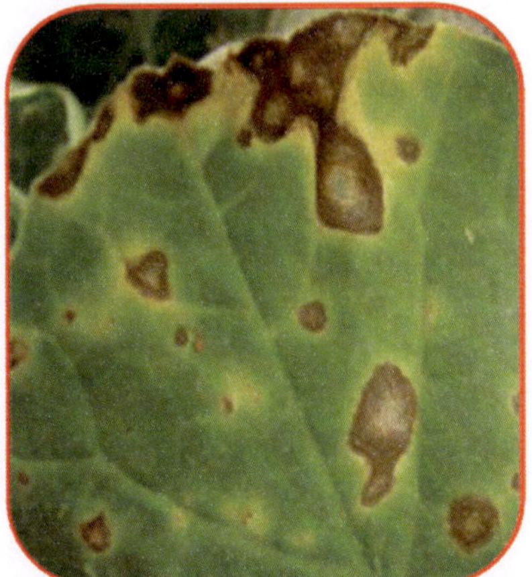

Fig. 1: Alternaria leaf blight

Fig. 2: Sclerotinia rot

Plate 14: Photograph showing symptoms of major diseases of Sunflower

15

Cotton Crop Diseases & Management

Name of Diseases	Causal Organism
Anthracnose	*Colletotrichum capsici*
Vascular wilt	*Fusarium oxysporum* f.sp. *vasinfectum* (Akt) Synedar & Hansen
Black arm	*Xanthomonas axonopodis* pv. *malvacearum* (Smith) Dye
Leaf blight	*Alternaria macrospora* Zimm
Leaf Curl	*Cotton leaf curl virus*

Anthracnose

Diagnostic Symptoms

- Pathogen infects the seedlings and produces small reddish circular spots on the cotyledons and primary leaves.
- Lesions develop on the collar region, stem may be girdled, causing seedling to wilt and die.
- In mature plants, the fungus attacks the stem, leading to stem splitting and shredding of bark.
- Most common symptom is boll spotting. Small water soaked, circular, reddish brown depressed spots appear on the bolls.
- The lint is stained to yellow or brown, becomes a solid brittle mass of fiber.The infected bolls cease to grow and burst and dry up prematurely(Fig-2).

Etiology

Scientific Classification

Kingdom : Fungi

Division : Eumycota

Sub- division :Deuteromycotina

Class : Coelomycetes

Order : Melanconiales

Family : Melanconiaceae

Genus : *Colletotrichum*

Species : *capsici*

- Pathogen: *Colletotrichum capsici.*
- Pathogen forms large number of acervuli on the infected parts.
- Conidiophores: are slightly curved, short and club shaped.
- Conidia are hyaline and falcate, borne single on the conidiophores.
- Setae: Numerous black colored and thick walled setae are also produced inacervulus.

Disease Cycle

- Pathogen survives as dormant mycelium in the seed or as conidia on the surface of seeds for about a year.
- Pathogen also perpetuates on the rotten bolls and other plant debris in the soil.
- Secondary spread is by air-borne conidia.

Epidemiology

- Prolonged rainfall at the time of boll formation.
- Close planting.

Management

- Remove and burn the infected plant debris and bolls in the soil.
- Rogue out the weed hosts.
- Treat the delinted seeds with Carbendazim or Thiram or Captan @ 2g/kg seed.
- Foliar spray with Mancozeb (0.2%) or Copper oxychloride (0.3%) or Carbendazim (0.1%) at boll formation stage of crop, if required repeat after 15 days.

Vascular Wilt

Diagnostic Symptoms

- Disease affects the crop at all stages.

Seedling stage

- Cotyledons of the seedling turn yellow and then brown.
- Base of petiole shows brown ring, followed by wilting and drying of the seedlings.

Young and grown up plants

- Symptoms start from the older leaves at the base, followed by younger ones towards the top, finally involving the branches and the whole plant.
- Initial symptom is yellowing of edges of leaves and area around the veins i.e. discoloration starts from the margin and spreads towards the midrib.
- The leaves loose their turgidity, gradually turn brown, droop and finally drop off.
- Defoliation or wilting may be complete leaving the stem alone standing in the field.
- Sometimes partial wilting occurs; where in only one portion of the plant is affected, the other remaining free.
- Browning or blackening of vascular tissues is the other important symptom.
- Black streaks or stripes may be seen extending upwards to the branches and downwards to lateral roots.
- In severe cases, discolouration may extend throughout the plant starting from roots extending to stem, leaves and even bolls.
- In transverse section, discoloured ring is seen in the woody tissues of stem.
- Plants affected later in the season are stunted with fewer bolls which are very small and open before they mature(Fig-1).

Etiology

Scientific Classification

Kingdom : Fungi

Division : Eumycota

Sub- division :Deuteromycotina

Class : Hyphomycetes

Order : Tuberculariales

Family : Tuberculariaceae

Genus : *Fusarium*

Species : *oxysporum* f. sp.*vasinfectum*

- Pathogen:*Fusarium oxysporum* f.sp. *vasinfectum.*
- Mycelium: white to grayish white or bluish purple, often form a mat on the collar region of the stem near the ground level.
- Hyphae: both inter as well as intracellular .
- Conidiophores: produced in sporodochia ,sometimes on the mycelium, vertically branched.
- Conidia: macrocondia, microconidia and chlamydospores.
- Macroconidia: 1 to 5 septate, hyaline, thin walled, falcate with inoculum ends, 40-50x3-4.5µm.
- Microconidia: hyaline, thin walled, spherical or elliptical, unicellular,measuring 5-12x 2-3.5 µm.
- Chlamydospores: dark coloured and thick walled, terminal or intercalary.

Disease Cycle

- It is a soil inhabitant and has good saprophytic ability.
- Pathogen can survive in soil as saprophyte and as chlamydospores for many years.
- Pathogen is both externally and internally seed-borne.
- Primary infection is mainly through conidia produced from dormant hyphae and chlamydospores in the soil.

- Secondary infection in case of this disease is rare as the conidia produced seldom succeed in establishing aerial infection on the plant.

Epidemiology

- Soil temperature – 20-30°C.
- Soil humidity- 40 to 70%.
- pH=5.3.
- Hot and dry periods followed by rains.
- Heavy black soils with an alkaline reaction.
- Increased doses of nitrogenous fertilizers.

Management

- Deep summer ploughing during June-July.
- Grow disease resistant varieties like Varalakshmi, Vijay Pratap, Jayadhar and Verum.
- Soak seed in 40-50 ppm streptocyclin solution before sowing.
- Treat the acid delinted seeds with Carboxin or Carbendazim at 2 g/kg.
- Foliar spraye with a mixture of Streptocyclin(Streptycyclin sulphate 90%+Tetracyclin hydrochloride 10%SP)on the appearance of field symptoms,if required repeat 10-15 days interval.
- Remove and burn the infected plant debris in the soil.
- Apply heavy doses of farm yard manure or other organic manures.
- Follow mixed cropping with non-host plants.
- Spot drench with Carbendazim 1g/litre.
- Delay planting to the end of october.
- Avoid dense planting.
- Retain cotton residues on the surface for 60 days.
- Bare fallow rotation is best.
- Always practice good farm hygiene.

Black Arm/Angular leaf Spot

Diagnostic Symptoms

- Pathogen attacks all stages from seed to harvest.
- Small water soaked spots , angular lesions of 1 to 5 mm across the leaves and bracts, especially on the undersurface of leaves. Hence called angular leaf spot.
- Sometimes extensive dark green, water soaked lesions along the veins known as vein blight.
- Lesions dry and darken with age and leaves may be shed prematurely resulting in extensive defoliation.
- Black lesions on the stem which girdle and spread along the stem or branch known as black arm.
- Dark green, water soaked, greasy, circular lesions of 2 to 10mm across the bolls, especially at the base of the boll under the calyx crown. As the boll matures the lesions dry out and prevent normal boll opening. This phase of symptom is called as "Boll rot".
- The bacterium spreads inside the boll and lint gets stained yellow because of bacterial ooze and loses its appearance and market value. The pathogen also infects the seed and causes reduction in size and viability of the seeds (Fig-3).

Etiology

Scientific Classification

Kingdom : Prokaryotae

Division : Gracilicutes

Class : Proteobacteria

Family : Psudomonadaceae

Genus : *Xanthomonas*

Species : *axonopodis* pv. *malvacearum*

- Pathogen: *Xanthomonas axonopodis* pv. *malvacearum.*
- Gram negative.

- Short rod with a single polar flagellum.
- Non-spore forming measures 1.0-1.2 X 0.7-0.9 μm.
- In culture, it forms yellow colony.

Disease Cycle

- Pathogen inoculums may either be present in the field on infected crop residues from a previous season or it may be introduced at planting within infected seed.
- Infected seeds are the source of primary inoculums.They carry the pathogen externally as well as internally.
- Plant debris is another source of inoculums.
- Lesions on cotyledons may be initiated by inoculums within the seed during germination.
- Inoculum from infected crop residues may be splashed onto the foliage and into the growing point of young seedlings where it can survive saprophytic ally on leaf surfaces.
- When environmental conditions are favorable the bacteria enter the plant via the stomata or wounds.
- As lesions develop pathogen exude out onto the leaf surface for further dispersal through wind driven rain.
- Pathogen is able to enter the seed when mature, open, blight-infected bolls are exposed to wet weather prior to harvest.

Epidemiology

- High atmospheric temperature – 30-40°C.
- Relative humidity – 85 per cent.
- Optimum soil temperature – 28°C.
- Early sowing.
- Delayed thinning.
- Poor tillage and late irrigation.

Management

- Grow resistant varieties like CRH 71 and Sujatha.
- Delint the cotton seeds with concentrated sulphuric acid at 100ml/kg of seed.
- Soak the seeds in Streptomycin sulphate (0.1%) overnight or treat the delinted seeds with Carboxin or Oxycarboxin at 2 g/kg.
- Foliar spraywith Streptomycin sulphate_+Tetracycline mixture 100g along with Copper oxychloride at 1.25 kg/ha.
- Remove and destroy the infected plant debris.
- Rogue out the volunteer cotton plants and weed hosts.
- Follow crop rotation with non-host crops.
- Early earthing up with potash.

Model Question Paper

A. Objective Type Questions

a. Choose the correct answer from the following

(1) *Colletotrichum capsici* belongs to order.

a. Melanconiales b. Peronosporales

c. Hyphomycetales d. Moniliales

(2) Anthracnose of cotton is caused by :

a. *Colletotrichum capsici* b. *Rhizoctonia bataticola*

c. *Sphacelotheca sorghi* d. *Sphacelotheca cruenta*

(3) The Angular leaf spot is caused by

a. *Xanthomonas axonopodis* pv. *malvacearum*

b. *Sphacelotheca cruenta*

c. *Rhizoctonia bataticola*

d. None of the above

(4) Black lesions on the stem which girdle and spread along the stem or branch of cotton known as.

a. Black arm
b. Angular leaf spot
c. Anthracnose
d. None

(5) Infected seeds are the source of primary inoculums and they carry the pathogen.

a. Externally
b. Internally
c. Externally as well as Internally
d. None

(6) *Xanthomonas axonopodis* pv. *malvacearum* is a type of bacteria

a. Gram negative
b. Gram positive
c. Gram neutral
d. None

(7) The pathogen *Fusarium oxysporum* f.sp. *vasinfectum*produces highly resistant surviving structure in soilcalled

a. Pseudothecia
b. Sclerotia
c. Chlaymydospore
d. Cleistothecia

(8) Fusarial wilt of cotton is caused by

a. *Ascochyta rabiei*
b. *Rhizoctonia bataticola*
c. *Fusarium oxysporum* f.sp. *vasinfectum*
d. *Sphacelotheca cruenta*

(9) *Fusarium oxysporum* f.sp. *vasinfectum produces*

a. Mcroconidia
b. Microconidia
c Chlaymydospores
d. All

(10) *Fusarium oxysporum f.sp. vasinfectum* belongs to order.

a. Tuberculariales
b. Peronosporales
c. Hyphomycetales
d. Moniliales

(a) Answer

S. No	Answer	S. No	Answer
1	a. Melanconiales	6	a. Gram negative
2	a. *Colletotrichum capsici*	7	c. Chlaymydospore
3	a. *Xanthomonas axonopodis* pv. *mal-vacearum*	8	c. *F. oxysporum* f.sp. *vasinfectum*
4	a. Black arm	9	d. All
5	c. Externally as well as Internally	10	a. Tuberculariales

(B) Fill in the blanks with suitable word (s)

(1) *Xanthomonas axonopodis* pv. *malvacearum* is a cotton....................... bacteria

(2) Anthracnose of cotton is caused by ....

(3) Numerous black colored and thick walled ----- -----are also produced in acervulus of *Colletotrichum capcici* causing anthracnose of cotton.

(4) Chlamydospore is the resting spore of the fungus ....

(5) *F oxysporum*f.sp. *vasinfectum* belongs to division--------------------

(6) *Colletotrichum capcici* belongs to order ------------------------

b. Answer

Sl.No	Answer	Sl.No	Answer
1	*Gram negative*	4	*Fusarium oxysporum* f.sp. *vasinfectum*
2	*Colletotrichum capsici*	5	Eumycota
3	Setae	6	Melanconiales

c. State whether the following statements are True or False

(1) *Xanthomonas axonopodis* pv. *malvacearum* forms yellow colony in culture.

(2) Foliar spray with Streptomycin sulphate_+Tetracycline mixture 100g along with Copper oxychloride at 1.25 kg/ha manage Angular leaf spot of cotton.

(3) Black lesions on the stem of cotton infected with *Xanthomonas axonopodis* pv. *malvacearum* which girdle and spread along the stem or branch known as black arm.

(4) The fungus *Fusarium oxysporum* f.sp. *vasinfectum*produces three types of conidia like macroconidia, microconidia and chlamydospores.

(5) Wilt disease of cotton can be managed by soil application of Trichoderma.

(6) An extensive dark green, water soaked lesions seen along the veins of cotton due to infection of *Xanthomonas axonopodis* pv. *malvacearum*known as vein blight.

(7) A small water soaked spots , angular lesions of 1 to 5 mm across the leaves and bracts, especially on the undersurface of leaves due to infection of *F oxysporum* f.sp. *vasinfectum* called angular leaf spot.

(8) *Xanthomonas axonopodis* pv. *malvacearum*is a gram negative, short rod shaped bacteria with a single polar flagellum.

(9) *Fusarium oxysporum* f.sp. *vasinfectum*is a soil inhabitant fungi having good saprophytic ability.

10. Yellowing of edges of leaves and area around the veins i.e. discoloration starts from the margin and spreads towards the midrib is a initial symptom of Fusarium wilt disease of cotton.

c. Answer

S. No	Answer	S. No	Answer
1	True	6	True
2	True	7	False
3	True	8	True
4	True	9	True
5	True	10	True

B. Descriptive Questions

a. Long answer questions

1. Describe in detail the occurrence, importance, symptoms, etiology, disease cycle and management of wilt of cotton.
2. Give the diagrammatic representation of disease cycle of the following diseases of cotton

   i. Wilt　　ii. Anthracnose

   iii. Bacterial blight　　iv. Leaf curl

3 Briefly describe the diagnostic symptoms and integrated management of any two of the following cotton diseases.

(i) Collar rot　　(ii) Bacterial blight

(iii) Wilt

4. Illustrate the Angular leaf spot of cotton in the following headings:
   (i) Pathogen (ii) Symptoms
   (ii) Disease cycle (iv) Disease management
5. Describe in detail the most distinguishing symptoms, causal organism, disease cycle and management of para wilt of cotton.

b. Short answer questions

1. Compare between symptom of angular leaf spot and black arm of cotton giving suitable measures for their management.
2. Give the disease cycle and predisposition of bacterial blight of cotton.
3. Write an integrated management scheduele of angular leaf spot of cotton.
4. Explain the disease cycle and epidemiological factor of vascular wilt of cotton.
5. Write systemic position and etiology of pathogen causing anthracnose of cotton.

c. Very short answer questions

1. Suggest some disease resistant varieties of cotton against vascular wilt disease.
2. Mention the name the resting structures by the following pathogens :
   (a) *Fusarium oxysporum f.* sp. *vasinfectum*
   (b) *Alternaria macrospora*
   (c) *Colletotrichum capsici*
   (d) *Rhizoctonia solani*
3. List some important diseases and their causal organisms of cotton.
4. Distinguish between fungal wilt and bacterial blight of cotton.
5. Which type of disease is can be effectively managed by crop rotation.

Fig. 1: Fusarium wilt

Fig. 2: Anthracnose

Fig. 3: Bacterial blight

Plate-15: Photograph showing symptoms of major diseases of cotton

16

Citrus Crop Diseases & Management

Name of Disease	Causal organism
Canker	*Xanthomonas campestris pv citri*
Gummosis	*Phytophthora parasitica, P. palmivora, P. citrophthora*
Scab/Verucosis	*Elsinoe fawcett*
Tristeza or Quick decline	*Citrus tristeza virus*
Greening	*Liberobactor asiaticum*
Exocortis of scaly butt	Viroid

1. Gummosis.

Diagnosic Symptoms

- First symptoms are dark staining of bark which progresses into the wood.
- Bark at the base is destroyed resulting in girdling and finally death of the tree.
- Bark in such parts dries, shrinks and cracks and shreds in lengthwise vertical strips.
- Later profuse exudation of gum from the bark of the trunk(characteristic symptom).
- The infection spreads both upward to the leaves and even fruits and downward to the roots causing fibrous root rot.
- The fibrous root rot destroys a major portion of the root system.
- However, the trunck and root infections may usually result in ultimate drying of the tree (Fig-1).

Etiology

Scientific Classification

Kingdom : Fungi

Division : Eumycota

Class : Oomycetes

Order : Pythiaceae

Family : Pyhiaceae

Genus : *Phytophthora*

Species : *palmivora/parasitica*

- Pathogen:*Phytophthora parasitica, P. palmivora, P. citrophthora*
- *P. palmivora* is the main cause of the disease in southern parts of our country although *P. parasitica* is also common in Karnataka.*P. parasitica* is found to cause the disease in Assam.

Character	***P. palmivora***	***P. parasitica***
Hyphae	Intercellular with haustoria and often swollen at regular intervals	Tough and irrregular
Sporangipphores	Simple or branched	Slender, irregulary os sympodically branched
Sporangia	Obpyriform and always terminal	Lateral or intercalary, broadly ovoid, obpyriform to spherical,papillate
Oospores	Spherical, thickwalled and produce secondary sporangium on germination	Apleurotic

Disease Cycle

- Pathogen survives in soil and plant debris.
- Primary source of infection is zoospores splashed on to the tree trunk.
- Secondary infection through spores that may be transported in rain or irrigation to the roots of the trees.
- They germinate and enter the root tip resulting in rot of the entire root let, later extending to the rest of the root

Epidemiology

- Prolonged contact of trunk with water as in flood irrigation; water logged areas and heavy soils.
- Optimum temperature for disease development -25-28^{0}C
- Soil pH-5.4-7.5

Management

- Selection of proper site with adequate drainage.
- Provision of an inner ring about 45 cm around the tree trunk to prevent moist soil.
- Avoid irrigation water from coming in direct contact with the trunk.
- Avoid injuries to crown roots or base of stem during cultural operations.
- Use resistant sour orange rootstocks for propagating varieties.
- Painting Bordeaux paste or with $ZnSO4$, $CuSO4$, lime (5:1:4) to a height of about 60 cm above the ground level at least once a year.
- Scrape the diseased portion with a sharp knife.
- Protect the cut surface with Bordeaux paste followed by spraying of 0.3% fosetyl-AL reduces the spread.
- Soil drenching with 0.2% metalaxyl and 0.5% *Trichoderma viride* commercial formulation is also effective.

2. Canker

Diagnostic Symptoms

- The disease appears as raised , rough, corky brownish pustules with characteristic halo on leaves.
- Canker lesions on the fruit do not possess the yellow halo as on leaves.
- Several lesions on fruit may coalesce to form a patch. The crater-like appearance is more marked on fruits than on leaves.
- On fruits, canker lesions reduce market value.
- The plants are also stunted and fruit yields are reduced considerably (Fig-2).

Etiology

Scientific Classification

Kingdom : Prokaryotae

Division : Gracilicutes

Class : Proteobacteria

Family : Psudomonadaceae

Genus : *Xanthomonas*

Species : *campestris* pv. *citri*

- Pathogen is gram negative, non spore forming and aerobic bacteria.
- Rod shaped.
- Size: 1.5-2.0 x0.5-0.75 µm.
- Forms chains and capsules.
- Motile by one polar flagellum.
- Colony: circular, straw yellow colored, slightly raised, smooth and shining.
- There are at least three types of citrus canker, caused by different strains. Cancrosis A, Cancrosis B &Cancrosis C.
- Cancrosis A (The Asiatic type), caused by A-strain, is the most severe form of the disease. It is found in India.

Disease Cycle

- Pathogen survives and multiplies in infected leaves, twigs and fruits.
- It may also survive in crevices in bark tissues of citrus trees.
- Infection occurs when wind blown rain carries inoculums to uninfected plants.
- Bacteria enter the leaf through stomata and lenticels, along with water and colonize the intercellular spaces.
- Infection also occurs through wounds caused by Citrus leaf miner larvae (*Phyllocnistis citrella*).
- Infection occurs on young leaves, twigs and the fruits.

Epidemiology

- Free moisture for 20 minutes
- Favourable temperature- 20-30°C.

Management

- Select seedlings free from canker for planting in main field.
- Prune out and burn all canker infected twigs before monsoon.
- Maintain proper aeration by training and pruning .
- Sterilize tools and equipment between uses to prevent the spread of the disease.
- Do not work in the field when foliage is wet.
- Clean boot, clothes thoroughly when working between ifferent orchards.
- Use windbreaks between fields to avoid propagation.
- Monitor the trees of symptoms of the disease.
- Foliar spray with Streptocycline (1g) + Copper oxy chloride (30g) in 10 litres of water at fortnightly intervals .
- SprayStreptocycline 50 to100 ppm solution repeatedly at an interval of 15 to20 days after the appearance of new growth.
- Cover the foliage and young fruits fully.

Model Question Paper

A. Objective Type Questions

a. Choose the correct answer from the following

(1) *Phytophthora palmivora* belongs to order.

a. Agonomycetales b. Peronosporales

c. Melanconiales d. Moniliales

(2) Citrus canker is caused by

a. *Xanthomonas campestris pv citri* b. *Rhizoctonia bataticola*

c. *Sphacelotheca sorghi* d. *Sphacelotheca cruenta*

(3) Gummosis of citrus is caused by:

a. *Phytophthora parasitica* b. *Oidium mangiferae*

c. *Sphacelotheca sorghi* d. *Sphacelotheca cruenta*

(4) Primary source of infection in gummosis of citrus is

a. Zoospore b. Aplanospore

c. Ascospore d. Basidiospore

(5) The main cause of the disease in southern parts of our country

a. *Phytophthora parasitica* *b. Phytophthora palmivora*

c. Phytophthora citrophthora d. All

(6) *Xanthomonas campestris* pv *citri* is a type of bacteria

a. Gram -ve b. Gram -ve

c. Both d. None

(7) Optimum temperature for development of powdery mildew of mango

a. 25-28^{0}C b. 25-40^{0}C.

c. 10-15^{0}C d. 15-20^{0}C.

(8) Metalaxyl fungicide used for control disease of citrus.

a. Powdery mildew b. Gummosis

c. Grey blight d. Citrus cnker

(9) The color of colony of *Xanthomonas campestris* pv *citri* is

a. yellow b. White

c. Green d. Pink

(10) The crater-like appearance is more marked on citrus fruits infected with disease

a. Powdery mildew b. Gummosis

c. Grey blight d. Citrus cnker

(11) Profuse exudation of gum from the bark of the citrus trunk is a characteristic symptom of disease

a. Powdery mildew b. Gummosis

c. Grey blight d. Citrus cnker

(12) *Xanthomonas campestris pv citri* belongs to the division

a. Gracilicutes b. Firmicutes

c. Tenericutes d. Mandicutes

Answer

S. No	Answer	S. No	Answer
1	b.Peronosporales	7	a. 25-28^{0}C
2	*a.Xanthomonas campestris pv citri*	8	b.Gummosis
3	*a.Phytophthora parasitica*	9	a. Yellow
4	a. Zoospore	10	d. Citrus cnker
5	*b Phytophthora palmivora*	11	b. Gummosis
6	a.Gram –ve	12	a. Gracilicutes

b. State whether the following statements are *True* or *False*

1. Secondary infection of anthracnose is mainly due to transfer of conidia through rain splash or wind driven rain water.
2. Hyphae of *Colletotrichum gloeosporioides* accumulate below the host cuticle and develop fruiting bodies called acervuli.
3. Fruits infected with Anthrcnose at mature stage carry the pathogen into storage and cause considerable loss during storage, transit and marketing.
4. Secondary spread of powdery mildew disease can occur thorugh air borne conidia produced in these infections.
5. Secondary infection of mango malformation occur either by air borne conidia or by conidia carried by eriophid mite *Aceria mangiferae.*
6. *Xanthomonas mangiferae- indicae* belong to family Psudomonadaceae.
7. Yellowish color of the colonies of *Xanthomonas mangiferae- indicae* is due to production of characteristic water in soluble, non-diffusible pigment.
8. Foliar spray of 0.02% Streptomycin followed by Copper oxychloride (0.3%) manages bacterial blight of mango.
9. Foliar sprays with wettable sulphur@ 0.3% or Tridemorph @ 0.1% control powdery mildew of mango.
10. Deficiency of Iron, Zinc and Copper can also cause the malformation.

Answer

S. No	Answer	S. No	Answer
1	True	6	True
2	True	7	True
3	True	8	True
4	True	9	True
5	True	10	True

B. Descriptive questions

a. Long answer questions

1. Explain in detail the, symptoms, etiology, disease cycle and management of citrus canker.
2. Give a thorough explanation of the signs and symptoms, disease cycle, epidemiology, and management of citrus gummosis.
3. List the main citrus diseases and the organism that causes them. Describe gummosis in terms of its symptoms, disease cycle and management.
4. Describe the gummosis in the following headings:

   (i) Pathogen (ii) Symptoms

   (ii) Disease cycle (iv) Disease management
5. Describe the citrus canker in the following headings:

   (i) Pathogen (ii) Symptoms

   (ii) Epidemiology (iv) Disease management

b.Short answer questions

1. Which species of citrus has been found to be the most susceptible to gummosis? Which pathogens cause this disease?
2. Write down the etiology of gummosis.
3. Which bacterial pathogen causes citrus canker? Give its systematic position.
4. Write down the integrated management practices of gummosis.
5. Describe the causal organisms and mode of penetration of citrus canker.

Fig. 1: Gummosis

Fig. 2: Citrus canker

Plate-16: Photograph showing symptoms of major diseases of citrus

17

Banana Crop Diseases & Management

S. No	Name of Disease	Causal Oranism
1.	Sigatoka	*Mycosphaerella gloeosporioides*
2.	Panama wilt	*Fusarium oxysporum f.sp cubense*
3.	Bacterial wilt	*Pseudomonas solanacearum / Burkholderia solanacearum*
4.	Bunchy top	*Banana bunchy top virus*
5.	Anthracnose	*Gloeosporium gloeosporioides*

1. Sigatoka

Diagnostic Symptoms

- Early symptoms appear on the young leaves.
- Small spindle shaped spots on foliage with greyish centreand yellowish halo running parallel to veins.
- Spots are mostly seen along the edge of the leaf with defined margin and possess dark brown to black margin. Spots coalesce and whole leaf blade dries up.
- On the upper surface of the spots, fructificationsof the fungus appear as black specks unevenly and individual bananas appear undersized and their flesh develops a buff pinkish colour, and store poorly (Fig. 1).

Etiology

Scientific Classification

Kingdom : Fungi

Division : Eumycota

Sub- division : Deuteromycotina

Class : Hyphomycetes

Order : Hyphomycetales (Moniliales)

Family : Dematiaceae

Genus : *Cercospora*

Species : *musae*

- Pathogen: *Mycosphaerella musicola (Cercospora musae*)**.**
- Mycelium is hyaline, septate and branched.
- Conidia are elongated, narrow and multiseptate.
- Perithecia are dark brown to black and ostiolate.
- Asci are oblong and clavate.
- Ascospores are hyaline, two celled, obtuse to ellipsoid.

Disease Cycle

- Pathogen survives on dry infected leaves on the field soil and primary infection takes place through ascospores in the infected plant debris.
- Secondary spread through wind borne conidia and ascospores.
- Infection takes place through stomata on the lower surface of young leaves.
- Surface moisture is necessary for release of both conidia and ascospores. Hence the disease is severe in moist weather.

Epidemiology

- High humidity, heavy dew and rainy weather with temp above 21^{0}C.
- Soils with poor drainage and low fertility favour the disease incidence.
- Thick planting, presence of weeds and increased number of suckers in a mat promote disease development.

Management

- Removal and destruction of the affected leaves.
- Prevent water accumulation around the plant and go for periodical weeding.
- Spray Zineb @0.25% or Copper oxy chloride @ 0.3% suspended in mineral oil.
- Spray chlorothalonil@0.2% or Carbendazim or Thiophanate methyl@ 0.1% or Mancozeb @ 0.25% along with spreading agent. Wetting agent such as teepol or sandovit added at the rate of 1ml/lit of water..

- Foliar spray of *Bacillus subtilis* @ 5 lit/ha. ·
- Spray Propiconazole @ 1 ml / litre of water for 3 to 5 times at 25 to 30 days interval.

2. Panama Wilt

Diagnostic Symptoms

- Symptoms are most clear on the plants that are at least five month old.
- Sudden wilting of the plant or individual leaves is a characteristic symptom of the disease.
- Yellowing of the lower most leaves starting from margin to midrib of the leaves.
- Yellowing extends upwards and finally heart leaf alone remains green for some time and it is also affected.
- The leaves break near the base and hang down around pseudostem.
- Discolored vascular strands varying from light yellow to dark brown are the distinguishing internal symptoms.
- Usually the discoloration manifests first in the outer or oldest leaf sheath and extends upto the pseudostem (Fig-2).

Etiology

Scientific Classification

Kingdom : Fungi

Division : Eumycota

Sub- division : Deuteromycotina

Class : Hyphomycetes

Order : Hyphomycetales (Moniliales)

Family : Dematiaceae

Genus : *Fusarium*

Species : *oxysporum* f. sp *cubense*

- Pathogen: *Fusarium oxysporum* f. sp *cubense*
- Mycelium is septate, hyaline and branched.

- Fungus produces micro, macro conidia and also chlamydospores.
- Micro conidia - Single celled or rarely one septate hyaline elliptical or oval.
- Macro conidia - Sickle shaped hyaline, 3-5 septate and tapering at both ends.
- Chalamydospores - Thick walled, spherical to oval, hyaline to slightly yellowish in color.

Disease Cycle

- The disease is soil borne.
- The pathogens survives in the soil mainly as chlamydospores and invades the soft roots or rhizomes, very often through the injuries.
- The pathogen can also survive in the diseased rhizomes and other plant parts saprophytically for a long period.
- Primary infection always takes place through injured roots.
- Deep wounding exposing the xylem help in easy infection.
- When the germ tube of chlaymydospores enters into the root, hyphae proceed internally along the root to the rhizome where they develop extensively in vascular tissues before passing up the vascular system into the pseudostem and bthe older leaf petioles.
- The secondary infection takes place by means of conidia.
- The disease can easily spread by contact of root system of adjacent healthy plants with conidia released by the diseased plants.
- Flood also help in the local dispersal of the secondary inoculums.

Epidemiology

- The disease is favoured by high soil moisture, light soils , bad drainage and the strong development of new roots.
- Optimem temperature-20-30^{0}C

Management

- Use of disease free suckers for planting.
- The planting material should be dipped in allyl alcohol.

- Dipping of suckers in Carbendazim (0.1%) solution before planting.
- Soil drench and corm injection with 0.1% Carbendazim or 0.01% Vapam.
- Infected plant should be eradicated by weed killer.
- Resistant cultivar should be cultivated. Cavendesh, Poovan, Moongil, Peyladen, Rajabale etc.
- Avoid ill drained soils, and prefer slightly alkaline soils (7-7.5 pH) for cultivation.
- Flood fallowing for 6 to 24 months or crop rotation with rice
- Application of lime (1-2 kg/pit) to the infected pits after chopping of the plants parts.

2. Bacteiral Wilt

Diagnostic Symptoms

- Symptoms start on rapidly growing young plants.
- The youngest three to four leaves turn pale green or yellow and collapse near the junction of lamina and petiole.
- Characteristic discoloration of vascular strands, wilting and blackening ofsuckers.
- Vascular discolouration (pale yellow to dark brown or bluish black) is concentrated near the centre of the pseudostem, becoming less apparent on the periphery.
- Greyish brown bacterial ooze is seen when the pseudostem of affected plant is cut transversely.
- A firm brown dry rot is found within fruits of infected plants (characteristic symptom) (Fig-3).

Etiology

Scientific Classification

Kingdom : Prokaryotae

Division : Gracilicutes

Class : Proteobacteria

Family : Psudomonadaceae

Genus : *Ralstonia*

Species : *solanacearum*

- Pathogen: *Pseudomonas solanacearum / Burkholderia solanacearum.*
- Moko disease is caused by race 2 of *Ralstonia solanacearum* which infects *Musa* and *Heliconia.*
- It is rod shaped, gram negative bacterium with one polar flagellum.

Disease cycle

- The pathogen survives in crop residues left in the field.
- Primary infection occurs when the bacteria enter the roots and rhizomes via wounds, caused by nematodes and farm implements.
- Secondary infections from plant to plants are caused by insects that transmit the bacterial oozes form infected male flowers to healthy flowers.

Epidemiology

- Warm and moist weather favours disease development.

Management

- Eradicate infected plant.
- Expose soil to direct sunlight.
- Use of clean planting material.
- Fallowing and crop rotation is advisable.
- Disinfestation of tools with formaldehyde diluted with water in 1:3 ratio.
- Crop rotation (3 years rotation with sugarcane or rice) & providing good drainage.
- Fumigation of infected site with Methyl Bromide or chloropicrin.
- Spray systemic insecticide to prevent transmission of disease to healthy plants.
- Biocontrol with *Pseudomonas fluorescens.*

7. Bunchy Top

Diagnostic Symptoms

- Dark broken bands of green tissues on the veins, leaves and petioles.
- Plants are extremely stunted.
- Leaves are reduced in size marginal chlorosis and curling.
- Leaves upright and become brittle.
- Many leaves are crowded at the top. Branches size will very small. If infected earlier no bunch will be produced (Fig-4).

Transmission

- Pathogen: *Banana bunchy top virus.*
- The virus is not transmitted by mechanical inoculation.
- The chief insect vector is *Pentalonia nigronervosa.*
- The aphid acquires the virus in a feeding period of 24 hours and transmits the virus to healthy plants to an infection feeding period of one and a half years.
- The aphid retains the virus for a period of 13 days and the incubation period for the virus is a few hours to two days.

Perpetuation

- The virus perpetuates through infected plant suckers.

Management

- Adaptation of strict quarantine measures.
- Eradication of all infected suckers by spraying with kerosene or by injecting herbicide, 2, 4-D.
- Certified viru free suckers should be for planting.
- Control vector by spraying Methyl demoton 1 ml/l.or Monocrotophos, 2 ml/l.or Phosphomidon 1 ml / lit. or Injection of Monocrotophos 1 ml / plant (1 ml diluted in 4 ml). Infected plants are destroyed using 4ml of 2, 4, D (50g in 400 ml of water).

Model Question Paper

A. Objective Type Questions

a. Choose the correct answer from the following

(1) *Cercospora musae* belongs to order.

a. Agonomycetales b. Peronosporales

c. Melanconiales d. Moniliales

(2) Sigatoka of banana is caused by

a. *Mycosphaerella gloeosporioides* b. *Rhizoctonia bataticola*

c. *Sphacelotheca sorghi* d. *Sphacelotheca cruenta*

(3) Bacterial wilt of banana is caused by:

a. *Pseudomonas solanacearum* b. *Oidium mangiferae*

c. *Sphacelotheca sorghi* d. *Sphacelotheca cruenta*

(3) Panama wilt of banana is caused by:

a. *Fusarium oxysporum* f.sp *cubense* b. *Oidium mangiferae*

c. *Sphacelotheca sorghi* d. *Sphacelotheca cruenta*

(4) Perithecia are dark brown to black resting structure with ostiole formed in

a. *Cercosopra musae* b. *Oidium mangiferae*

c. *Phytophthora parasitica* d. *Sphacelotheca cruenta*

(5) The aphid retains the Banana bunchy top virus for a period of

a. 13 days *b.* 20 days

c. 25 days d. 30 days

(6) *Xanthomonas campestris* pv *citri* is a type of bacteria

a. Gram -ve b. Gram +ve

c. Both d. None

(7) Optimum temperature for development of Panama wilt of banana

a. 20-30^{0}C b. 30-40^{0}C.

c. 10-15^{0}C d. 40-45^{0}C.

(8) The fungicide Thiophanate methyl @ 0.1% used for control disease of banana.

a. Moko b. Cigatoka

c. Grey blight d. Bunchy top

(9) The color of colony of*Pseudomonas solanacearum* is

a. Yellow b. White

c. Green d. Pink

(10) Small spindle shaped spots on foliage with greyish centre and yellowish halo running parallel to veins is characteristic symptoms of

a. Powdery mildew b. Gummosis

c. Grey blight d. Sigatoka

(11) Moko disease is caused by which race of *Ralstonia solanacearum*

a. Race-1 b. Race-2

c. Race-3 d. Race-4

(12) *Ralstonia solanacearum* belongs to the division

a. Gracilicutes b. Firmicutes

c. Tenericutes d. Mandicutes

(13) The fungus *Cercospora musae* survives on dry infected leaves of banana on the field soil and primary infection takes place through in the infected plant debris.

a. Zoospore b. Aplanospore

c. Ascospore d. Basidiospore

(14) The pathogen *Fusarium oxysporum* f.sp *cubense* perennates in soil through its

a. Oospores b. Cleistothecium

c. Chlaymydospore d. Apothecium

(15) A firm brown dry rot is found within fruits of infected plants is characteristic symptom of

a. Powdery mildew b. Gummosis

c. Grey blight d. Moko

(16) The chief insect vector of bunchy top of banana is

a. *Pentalonia nigronervosa* b. *Bemesia tabaci*

c. *Aphid gossypii* d. *Acaria cajani*

Answer

S. No	Answer	S. No	Answer
1	c. Melanconiales	9	b. White
2	a. *Mycosphaerella gloeosporioides*	10	d. Sigatoka
3	a. *Fusarium oxysporum* f.sp cubense	11	b. Race-2
4	a. *Cercosopra musae*	12	a. Gracilicutes
5	*b.* 13 days	13	c. Ascospore
6	a. Gram -ve	14	c. Chlaymydospore
7	a. 20-30^{0}C	15	d. Moko
8	b. Cigatoka	16	a. *Pentalonia nigronervosa*

b. State whether the following statements are True or False

1. Secondary infections in bacteiral wilt of banana from plant to plants are caused by insects that transmit the bacterial oozes form infected male flowers to healthy flowers.
2. *Pseudomonas solanacearum*is a rod shaped gram negative bacterium with one polar flagellum.
3. *Fusarium oxysporum* f. sp *cubense* is soil borne in nature.
4. *Fusarium oxysporum* f. sp *cubense* produces micro, macro conidia and also chlamydospores.
5. Sudden wilting of the plant or individual leaves is a characteristic symptom of the panama wilt disease of banana.
6. Small spindle shaped spots on foliage with greyish centreand yellowish halo running parallel to veins are characteristic symptom of sigatoka disease of banana.
7. *Mycosphaerella musicola produces* dark brown to black perithecia having ostiole.
8. Primary infection always takes place through injured roots in panama wilt disease of banana.

9. The *Banana bunchy top virus* is not transmitted by mechanical inoculation.
10. The *Banana bunchy top virus* perpetuates through infected plant suckers.

Answer

1.	True	6.	True
2.	True	7.	True
3.	True	8.	True
4.	True	9.	True
5.	True	10.	True

B. Descriptive Questions

a. Long answer questions

1. Explain in detail the occurrence, importance, symptoms, etiology, disease cycle and management of Sigatoka diseae of banana.
2. Write down the disease cycle and epidemiology of the following diseases of grape

   (i) Panama wilt (ii) Bacterial wilt

   (iiii) Sigatoka
3. Briefly describe the diagnostic symptoms and integrated management of any two of the following citrus diseases:

   (i) Bunchy top (ii) Sigatoka

   (iii)Panama wilt
4. Describe the panama wilt disease of banana in the following headings:

   (i) Pathogen (ii) Symptoms

   (ii) Disease cycle (iv) Management
5. Explain the bacterial wilt of banana in the following headings:

   (i) Symptoms (ii) Disease cycle

   (iii)Epidemiology (iv) Management
6. Explain in detail the, symptoms, transmission, perpetuation and management of bunchy top of banana.

b. Short answer questions

1. Write down the disease cycle of sigatoka disease of banana.
2. Write down the etiology of panama wilt of banana.
3. Which pathogen causes bacterial wilt of banana? Give its systematic position.
4. Write down the integrated management practices of panama wilt.
5. Describe the causal organisms and transmission of banana bunchy top.
6. Differentiate between bacterial wilt and panama wilt of banana.

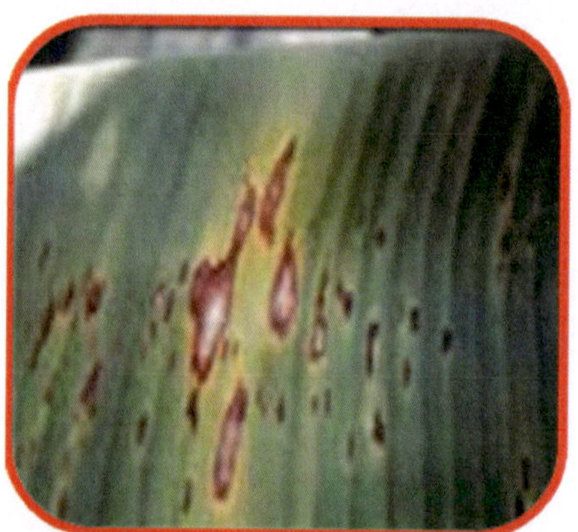

Fig. 1: Yellow sigatoka

Fig. 2: Panama wilt

Fig. 3: Bacterial wilt

Fig. 4: Bunchy top

Plate-17: Photograph showing symptoms of major diseases of banana

18

Papaya Crop Diseases & Management

Name of Disease	Causal organism
Foot rot	*Pythium aphanidermatum*
Mosaic	Papaya mosaic virus
Leaf curl	Nicotiana virus 10
Ring spot	*Papaya ring spot virus*
Powdery mildew	*Oidium caricae*
Anthracnose	*Colletotrichum gleoesporiodes*

1. Stem rot / Foot rot/Collar rot

Diagnostic Symptoms

- The appearance of water-soaked spots on the stem close to ground level is what characterizes it.
- The tissues degrade as a result of the fast enlargement of these patches, girdling the stem and turning it dark brown or black.
- Strong winds cause these afflicted plants to topple over and eventually die.
- If the disease attack is not too severe, the plants will still be stunted and only one side of the stem will decay.
- If fruit develops, it is deformed and withered.
- The plant eventually dies (Fig.1).

Etiology

- Pathogen: *Pythium aphanidermatum.*
- Mycelium is colorless, slender, coenocytic, profusely branching and rapidly growing mostly intracellularly.
- Sporangia terminal or intercalary produce a vesicle at the end of hyphal branch.
- Zoospores biflagellate and reniform.

Disease Cycle

- The pathogen perennates in soil through its oospores occurring in infected plant debris or most commonly, through mycelium.
- The perennating oospore or saprophytic mycelium occurring in soil serve as the source of primary infection.
- Germ tubes resulted in by the germination of oospores, or of zoospores or saprophytic mycelium when it comes in contact with the ground level part of the plant , perennate therein and cause primary infection.
- The zoospores produced as a result of primary infection get disseminated by water currents, reach the ground level parts of the healthy papaya plants, and cause infection thereupon.
- On the later part of growth phage the pathogen growing as a result of primary or secondary infection undergoes sexual course of its life and produces oospores.
- The latter perennate and serve as the source of primary infection for the next growing season.

Epidemiology

- Excessive rains and relatively high temperature (around 36^0C).
- Appears from June to August.
- Younger seedlings are more susceptible than older ones.
- Severity increases with intensity of rainfall.

Management

- Seedlings should be raised in well drained nursery area.
- Uproot the diseased seedlings and burn.
- Water logging in the field should be avoided.
- Affected pit should not be used for replanting.
- Seed treatment with captan@4g/kg seed or chlorothalonil@2g/kg seed (*R. solani*).
- Drench the base of stem with Copper oxy chloride @0.25% or metalaxyl@0.1% or Bordeaux mixture@1.2%.

2. Mosaic

Diagnostic Symptoms

- Causes leaf mosaic and stunting in papaya.
- Young seedlings in the greenhouse show vein-clearing and downward cupping of the leaves about 5 days after inoculation.
- A mottle or mosaic develops after 15-20 days.
- Symptoms appear on the young leaves of the plants.
- The leaves are reduced in size and show blister like patches of dark-green tissue, alternating with yellowish-green lamina.
- The leaf petiole is reduced in length and the top leaves assume an upright position (Fig.2).

Transmission and Favourable Conditions

- The papaya mosaic virus, which is aphid-borne and limited in host range to papaya and cucurbits.
- It is a mechanically transmissible virus that is linked to other viral diseases.

Management

- Raise papaya seedlings under insect-proof conditions.
- Plant disease free seedlings.
- Raise sorghum / maize as barrier crop before planting papaya.
- Rogue out affected plants immediately on noticing symptoms.
- Donot raise cucurbits around the field.
- Check transplants for aphids before planting.
- Reflective mulches such as silver colored plastic can deter aphids from feeding on plants.
- Sturdy plants can be sprayed with a strong jet of water to knock aphids from leaves.

3. Leaf Curl

Diagnostic Symptoms

- Curling, crinkling and distortion of leaves, reduction of leaf lamina, rolling of leaf margins inward and downward, thickening of veins.
- Leaves become leathery, brittle and distorted. Plants stunted. Affected plants does not produce flowers and fruits.
- Spread by whitefly *Bemisiatabaci* (Fig.3).

Transmission and Favourable conditions

- The virus can not be transferred mechanical means as in the case of mosaic disease.
- The virus readily transmitted through grafting and white fly (*Bemisia tabaci*).

Management

- Uproot the virus affected plants.
- Avoid growing tomato, tobacco near papaya.
- Control whitefly vector.
- Removal and destruction of the affected plants is the only control measure to reduce the spread of the disease.
- The field should be kept weed free. Tobacco, Tomato. Sunnhemp, Cape gooseberry, Chilli, Petunia, Datura Zinnia etc. should not be grown nearby papaya field.

4. Ring Spot

Diagnostic Symptoms

- Vein clearing, puckering and chlorophyll leaf tissues lobbing in.
- Margin and distal parts of leaves roll downward and inwards, mosaic mottling, dark green blisters, leaf distortion which result in shoe string system and stunting of plants.
- On fruits, circular concentric rings are produced. If affected earlier no fruit formation.
- Vectored by aphids *Aphis gossypii, A. craccivora* and also spreads to cucurbits not through seeds (Fig.4).

Transmission and Favourable Conditions

- Disease is aphid transmitted and aphids are more active during warmer conditions.
- PRSV is also easily transmitted via mechanical inoculation but there are no confirmed reports of PRSV transmission through seeds.

Management

- Use of yellow sticky strap to control of aphid vector.
- Use of resistant varieties.
- Early detection of infected plants and prompt removal can check the spread of the disease.
- Rogue out infected plants of papaya as early as possibleto avoid further infection within the field.
- Avoid taking mixed crop of tobacco, chilli, zinnea, tomato and gooseberry in papaya field or nearby.

Model Question

A. Objective Type Questions

a. Choose the correct answer from the following

(1) *Pythium aphanidermatum* belongs to

a. Pleosporales b. Peronosporales

c. Melanconiales d. Moniliales

(2) Foot rot of papaya is caused by

a. *Pythium aphanidermatum* b. *Rhizoctonia bataticola*

c. *Sphacelotheca sorghi* d. *Sphacelotheca cruenta*

(3) Leaf curl of papaya is caused by

a. Nicotiana virus 10 b. *Uncinula necator*

c. *Papaya ring spot virus* d. *Papaya mosaic virus*

(4) Mosaic of papaya is caused by

a. Papaya mosaic virus b. Nicotiana virus 10

c. Papaya ring spot virus d. All

(5) The pathogen *Pythium aphanidermatum* perennates in soil through its

a. Oospores
b. Cleistothecium
c. Pseudotheicum
d. Apothecium

(6) Papaya mosaic virus is transmitted by

a. Aphid
b. White fly
c Leaf hopper
d. Mite

(7) Leaf curl disease of papaya is transmitted by

a. Aphid
b. White fly
c. Leaf hopper
d. Mite

(8) Ring spot of papaya is transmitted by

a. Aphid
b. White fly
c. Leaf hopper
d. Mite

Answer

S. No	Answer	S. No	Answer
1.	b.Peronosporales	5	a. Oospores
2.	a. *Pythium aphanidermatum*	6	a. Aphid
3.	a. Nicotiana virus 10	7	b. White fly
4.	a. Papaya mosaic virus	8	a. Aphid

b. State whether the following statements are True or False

1. The appearance of water-soaked spots on the stem close to ground level is what characterizes collar rot of papaya.
2. *Pythium aphanidermatum* mycelium is colorless, slender, coenocytic, profusely branching and rapidly growing mostly intracellularly.
3. The papaya mosaic virus is aphid-borne.
4. Curling, crinkling and distortion of leaves, reduction of leaf lamina, rolling of leaf margins inward and downward, thickening of veins are characteristic symptom of leaf curl.
5. *Pythium aphanidermatum* sporangia are terminal or intercalary produce a vesicle at the end of hyphal branch.
6. Zoospores of *Pythium aphanidermatum* biflagellate and reniform.
7. The virus readily transmitted through white fly (*Bemisia tabaci*).

8. The pathogen *Pythium aphanidermatum* perennates in soil through its oospores occurring in infected plant debris .
9. The perennating oospore or saprophytic mycelium of *Pythium aphanidermatum* occurring in soil serve as the source of primary infection.
10. Younger seedlings of papaya are more susceptible to damping off than older ones.
11. Leaf curl disease isspread by whitefly *Bemisiatabaci.*
12. Ring spot disese of papaya is aphid transmitted.

Answer

S. No	Answer	S. No	Answer
1	True	7	True
2	True	8	True
3	True	9	True
4	True	10	True
5	True	11	True
6	True	12	True

B. Descriptive Questions

a. Long answer questions

1. What are the major fungal diseases that affect papaya plants, and how can they be managed effectively?
2. Explain the various viral diseases of papaya, with a focus on Papaya Ringspot Virus (PRSV), its symptoms, transmission, and the strategies to control its spread.
3. Describe the impact of bacterial wilt on papaya crops and discuss integrated disease management (IDM) strategies for managing bacterial wilt.
4. What is papaya leaf spot disease, and how can it be effectively controlled using cultural, biological, and chemical control measures?
5. How do soil-borne diseases like root rot affect papaya, and what are the best management practices to prevent and control these diseases?
6. Discuss the role of nematodes in papaya cultivation, the symptoms of nematode infestation, and effective strategies for managing nematode populations.
7. What is papaya mosaic disease, and what are the challenges in managing this disease in papaya orchards?

b. Short answer questions

1. What is the most common fungal disease affecting papaya plants?
2. Which bacterial disease causes leaf spots and fruit rot in papaya?
3. How can papaya mosaic virus be managed?
4. What is the key symptom of papaya dieback disease?
5. What cultural practice helps reduce the spread of papaya ringspot virus (PRSV)?
6. How can papaya root rot caused by Phytophthora be controlled?
7. What is a common management practice to prevent papaya anthracnose disease?

Fig. 1: Foot rot

Fig. 2: Mosaic

Fig. 3: Leaf curl

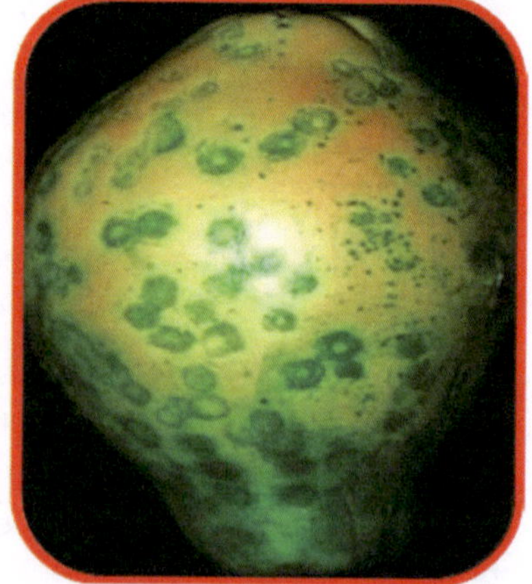

Fig. 4: Ring spot

Plate-18: Photograph showing symptoms of major diseases of papaya

19

Pomegranate Crop Diseases & Management

Name of Disease	**Causal organism**
Bacterial blight	*Xanthomonas axonopodis pv. punicae*
Wilt	*Ceratocystis fimbriata*
Cercospora leaf and fruit spots	*Cercospora punicae*
Anthranose	*Colletotrichum gloeosporioides*
Sun scald	High temperature along with excessive light

1) Bacterial Blight

Diagnostic Symptoms

- Bacterial blight of pomegranate was first reported in India from Delhi in 1952.
- First water-soaked lesions develop on fruits within 2–3 days and appear as spots.
- Initially, spots are black and round and surrounded by bacterial ooze.
- Under favorable conditions, spots enlarge to become raised, dark brown lesions with indefinite margins that cause the fruit to crack.
- Disease may cause up to 90% yield reduction (Fig.1).

Etiology

Domain : Bacteria

Phylum : Proteobacteria

Class : Gammaproteobacteria

Order : Xanthomonadales

Family : Xanthomonadaceae

Genus : *Xanthomonas*

Species : *citri pv. punicae*

1. Pathogen*: Xanthomonas axonopodis pv. punicae.*
2. Gram-negative, aerobic, short rod-shaped bacterium.
3. On common culture media, such as nutritional agar (NA), pale yellow, domed, round colonies with a diameter of 1-2 mm are seen two to three days after incubation.
4. Bacteria is catalase positive, gram and oxidase negative in addition to having an oxidative metabolism.
5. On YDC medium, it forms pale yellow, mucoid colonies; on 0.1% triphenyl tetrazolium chloride (TTC), it does not grow.
6. The cells are rod-shaped and 0.4-1.0 µm broad and 1.2-3.0 µm long, with a single polar flagellum that allows for movement.

Disease Cycle

- Bacterial cells are capable of surviving in soil for >120 days and also survive in fallen leaves during the off-season.
- Infected fruit and twigs are potential sources of primary inoculum.
- Secondary spread of bacterium is mainly through rain and spray splashes, irrigation water, pruning tools, humans, and insect vectors.
- Pathogen enter through wounds and natural openings.
- Disease buildup is rapid from July to September.
- Severity increases during June and July and reaches a maximum in September and October and then declines.

Epidemiology

- High temperatures and low humidity.

7. Temperature of 25–30°C is ideal for growth (ideally 27°C).

- Thermal death point is about 52°C.

Integrated Management

- Use of disease free planting material.
- Follow clean cultivation and orchard sanitation.

- Mass eradication measures needed to be taken up throughout the country with the help of department, university, growers to root out the disease from country.
- Crop should be taken only after 3-4 years after planting.
- Pruning should be done in August to September in order to get lesser severity and with less number of sprays.
- Foliar sprays with Streptocycline (500 ppm) + Copper oxy chloride (2000 ppm) or Sterptocycline (500 ppm) and K-cycline (500 ppm) , if required with repeat after 15 days.

Model Question Paper

A, Objective Type Questions

a. Choose the correct answer from the following

(1) *Xanthomonas axonopodis pv. punicae* belongs to order.

a. Exobasidiales b. Peronosporales

c. Xanthomonadales d. Moniliales

(2) Bacterial blight of pomegranate is caused by

a. *Xanthomonas axonopodis pv. punicae*

b. *Erwinia carotobvora*

c. *Pseudomonas axonopodis pv. punicae*

d. *Bacillus subtilis*

(3) Bacterial blight of pomegranate was first reported from

a. India b. Pakistan

c. Bangladesh d. Bhutan

(4) Bacterial blight of pomegranate was reported from which state of India

a. Delhi b. Madhya Pradesh

c. Uttar Pradesh d. Himachal Pradesh

(5) What color colones do *Xanthomonas axonopodis pv. punicae* develop on YDC media?

a. Yellow b. Red

c. Brown d. White

(6) Ideal temperature for growth of *Xanthomonas axonopodis pv. punicae is*

a. 17°C b. 27°C

c. 10°C d. 37°C

(7) *Xanthomonas axonopodis pv. punicae* bacteria is

a. Gram-negative b. Aerobic

c. Rod-shaped d. All

Answer

S. No	Answer	S. No	Answer
1	c. Xanthomonadales	5	a.Yellow
2	a. *Xanthomonas axonopodis pv. punicae*	6	b. 27°C
3	a. India	7	d.All
4	a. Delhi		

B. State whether the following statements are True or False

1. Bacterial blight of pomegranate was first reported in India from Delhi in 1952.
2. The pathogen *Xanthomonas axonopodis pv. punicae* isgram-negative, aerobic short rod-shaped bacterium.
3. The pathogen *Xanthomonas axonopodis pv. punicae* have an oxidative metabolism in addition to being gram, oxidase, and catalase negative.
4. *Xanthomonas axonopodis pv. punicae* produces pale yellow, mucoid colonies on YDC medium .
5. *Xanthomonas axonopodis* pv. *punicae* does not grow on 0.1% triphenyl tetrazolium chloride (TTC).
6. *Xanthomonas axonopodis pv. punicae*cells are rod-shaped and 0.4-1.0 µm broad and 1.2-3.0 µm long, with a single polar flagellum that allows for movement.
7. Fruit cracks as a result of bacterial blight patches growing into elevated, dark brown lesions with infinite margins under optimum conditions in pomegranate.
8. Pomegranate bacterial blight main inoculums may originate from infected fruit and twigs.

9. The primary means of secondary transmission for *Xanthomonas axonopodis* pv. *punica* are insect vectors, irrigation water, pruning instruments, rain and spray splashes, and people.
10. Bacterial blight spreads rapidly between July and September in Pomegranate.
11. PCR-based early and reliable diagnostic techniques are available for detection and confirmation of the *Xanthomonas axonopodis pv. punicae* on pomegranate.

Answer

S. No	Answer	S. No	Answer
1	True	6	True
2	True	7	True
3	True	8	True
4	True	9	True
5	True	10	True
		11	True

B. Descriptive Questions

a. Long answer questions

1. Describe in detail the most distinguishing symptoms, causal organism, disease cycle, epidemiology and management of bacterial blight of pomegranate.
2. Explain in detail about symptoms, etiology, disease cycle and management of bacterial blight of pomegranate.
3. Describe the bacterial blight of pomegranate in the following headings:

   (i) Pathogen (ii) Symptoms

   (iii) Disease cycle (iv) Disease management

b. Short answer questions

1. Which fungal pathogen causes bacterial blight of pomegranate? Give its systematic position.
2. Write down the management practices of bacterial blight of pomegranate.
3. Describe the causal organisms and diagnostic symptoms of bacterial blight of pomegranate.

4. Why has pomegranate bacterial blight become such a significant disease in India recently?
5. Write down the etiology of *Xanthomonas axonopodis pv. punicae.*

Fig. 1: Bacterial blight

Plate-19: Photograph showing symptoms of major diseases of pomegranate

20

Apple Crop Diseases & Management

Name of Diseases	Causal Organism
Scab	*Venturia inaequalis* (Cooke) Wint
Powdery mildew	*Podosphaera leucotricha* (Ell. And Ev.)
Fire blight	*Erwinia amylovora* (Burrill) Winslow et al
Crown gall	*Agrobacterium tumefaciens* (E.F. Sm. & Towns) Conn

1). Scab

Diagnostic Symptoms

- Symptoms generally noticed on leaves and fruits.
- Affected leaves become twisted or puckered and have black, circular spots on their upper surface.
- On the under surface of leaves, the spots are velvety and may coalesce to cover the whole leaf surface. Severely affected leaves may turn yellow and drop.
- Scab can also infect flower stems and cause flowers to drop.
- Lesions later become sunken and brown and may have spores around their margins.
- Infected fruit become distorted and may crack, allowing entry of secondary organisms(Figure-1).

Etiology

Scientific Classification

Kingdom : Fungi

Division : Ascomycota

Class : Loculoascomycetes

Order : Pleosporales

Family : Venturiaceae

Genus : *Venturia*

Species : *inaequalis*

- Pathogen: *Ventruia inaequalis* (*Spilocaea pomi* Fr.).
- Pathogen has got two distinct stages- the imperfect stage(*Spilocaea pomi*) stage on living plant parts and perfect (*Ventruia inaequalis*) on fallen dead leaves.
- Fungus produces ascocarps in a stroma in over wintered leavesin orchard floor.
- Pseudothecia are dark brown to black and spherical 90-150μm inn diameter, having short beak and a distinct ostiole having single celled bristles at the apex.
- Asci: are cylindrical, fasciculate short stipulate produced in large number in the pseudotheicum which vary in size from 55-75x6-12 μm.
- Ascospores: measure 11-15x5-7 μm which are yellowish green to unequally bicelled, with upper cell smaller and wider than the lower one.
- Conidia are produced on conidiophores whicharise from short erect, closely septate brown mycelium and are non septate or have a single septum. Conidia are produced on leaf surface and fruit lesions throughput the spring and summer and measure 12-12x6-9μm.
- Fungus is bipolar heterothallic.
- There are five races of the pathogen .Out of these two have been reported to occur in India.

Disease cycle

- Pathogen survives through perithecia in the soil debris.
- Soil inoculums is the source of primary infection.
- Secondary infections occurs by conidia through rain splash or wind.

Epidemiology

- High soil moisture.
- Temperature -25^0 C.
- Leaf wetness.

- Pathogen consists of weft of white hyphal threads and draw nutrition through the haustoria.
- Conidia: produced in chain on conidiophores,hyaline clear, without color, measure 20-38 × 12 μm, and contain distinct fibrosin bodies.
- Cleistothecia: ascocarp of pathogen, globose, black, partly embedded in mycelium.Appendages are long and stiff or short and tortuous.
- Asci: sac-like , enclosed in fruiting bodies ascocarps, contains a single ascus with eight ascospores, each of which is elliptical and measures 22-36 x 12-15 μm. Ascocarps are densely grouped together, measure 75-96 μm in diameter and have apical and basal appendages.

Disease Cycle

- Pathogen survives in the form of resting mycelium or encapsulated haustoria in the dormant terminal and lateral shoot buds and in blossom buds produced and infected the previous growing season.
- Abundant sporulation from overwintering shoots and secondary lesions on young foliage leads to a rapid build up of inoculum.
- Secondary spread occurs through wing borne conidia.
- Cleistotheica are produced on heavily infected shoot and leaves in mid summer but they are not regarded as an important inoculums source because the ascoposres do not germinate readily.

Epidemiology

- High relative humidity –greater than 70%.
- Temperature – 10 and 25°C.
- Youngest leaves are the most susceptible.

Management

- Plant resistant varieties.
- Plant crops with sufficient spacing to allow for good ventilation.
- Remove infect leaves when the first spots appear.
- Donot touch healthy plants after touching infected plants.
- Fertilize with a balanced nutrient supply.
- Remove plant residue after harvest.

Management

- Use resistant or tolerant varieties.
- Collect and burn of fallen leaves, shoots and fruits.
- Ensure pruning method that allow for more air circulation.
- Avoid getting foliage wet when watering.
- Resistant varieties: Ashwini, Ambika, Angeles, American pride, Surabh
- Sprays with wettable Difenoconazole @0.1% or Captan @0.25% c Carbendazim@0.1% at15 days interval.

2) Powdery Mildew

Diagnostic Symptoms

- Symptoms produced on young shoots, leaves, blossoms, and fruit.
- White or grey patches appears on the leaves.
- Leaf fold, twist and cracking may takes place.
- Buds get infected and may be killed.
- Both leaves and twigs may be covered with white powdery mass.
- Defoliation of leaves and deformation of fruits takes place. (Figure-2)

Etiology

Scientific Classification

Kingdom : Fungi

Division : Eumycota

Phyllum: Ascomycota

Class : Ascomycetes

Order : Erysiphales

Family : Erysiphaceae

Genus : *Oidium*

Species : farinosum

- Pathogen*: Podosphaera leucotricha* (*Oidium farinosum*).
- Obligate parasite.

- Foliar spray with Carbendazim 50% WP 0.2% or benomyl 50% WP @ 0.05% or Thiophanate methyl 70% WP @ 0.1% or Calixin 0.1% or Karathane @ 0.2%, if required repeat after 15 days.
- Monitor fields regularly to assess the incidence of a disease or pest.

3) Fire Blight

Erwinia amylovora pv.pyri

Diagnostic Symptoms

- Term 'fire blight' sums up the symptoms as if scorched by fire. Primary infection occurs on flowers causing 'blossom blight' , followed by the more serious 'shoot blight'.
- Infected blossoms become brown, later turning black and giving a burnt appearance.
- Flower become water soaked and brownish to black in color,may fall or remain hanging.
- Brwon black blotches appear on the leaves, blackening advances and leaves curl and hang.
- Watersprouts and terminal twigs wilt, die-back extends down the water sprout and twigs.
- Canker appear on branches which may girdle the branch and trunk to cause collar rot.
- Infected fruit becomes water soaked and brown to black in color. Brown ooze may be seen on the surface of infected parts under humid conditions. (Figure-3).

Etiology

Scientific Classification

Kingdom : Bacteria

Phyllum : Protobacteria

Class : Gammaproteobacteria

Order : Enterobacteriales

Family : Enterobacteriaceae

Genus : *Erwinia*

Species : *amylovora* pv.*pyri*

- The pathogen is of historical importance. It has the distinction of being the first bacterium of plants shown to be a pathogen.
- The bacterium can be a plant pathogen was given by T.J.Burrill in 1878.
- *Erwinia amyovora* is also the first pathogen to be associated with an inscet vector.
- Rod shaped, filamentous bacterium.
- Gram negative.
- Size 1.1-3.0 μm x 0.5-1.2 μm.
- Motile by many peritrichous flagella, aerobic or facultative anaerobic, non sporulating.
- Colonies of virulent isolates are small, round and white with typical glistening shine.
- Bacterium requires nicotininc acid as a growth factor.

Disease Cycle

- Pathogen survives primarily in cankers, blighted twigs and bud of symptomless twigs and serve as primary inoculums for infection next spring.
- During spring and early summer, infected parts reactivate and produce bacterial ooze consisting of millions of bacterial cells and easily transported to blossoms by insects such as flies, ants, and beetles.
- Bacteria can then be spread very efficiently from blossom to blossom by honey bees causes blossom blight.
- Once blossoms are infected, the bacteria can quickly spread and cause secondary infection into shoots and branches, which can be spread by rain, insects, and contaminated pruning tools.Secondary infections may continue to occur throughout the growing season.
- Secondary infections multiply the disease and finally result in infections of the tree trunk and the root stock.

Epidemiology

- Temperature- 21 to 23^{0}C.
- Warm cloudy and rainy weather.

Integrated Management

- Grow resistant varieties, if available.
- Monitor orchards regularly for signs of the disease.
- Prune out infected branches and burn them , preferably by the end of the winter.
- Donot apply excess nitrogen on the trees.
- Shave off the cankers and treat with Bordeaux paste.
- Foliar spray with Bordeaux mixture (0.1%) along with Streptocylin or Terramycin.
- Regulate soil fertility to reduce succulence of new growth.

4) Crown Gall

Agrobacterium tumefaciens

Diagnostic Symptoms

- Small tumours or galls develop in the wound just below the soil surface. They enlarge rapidly with outer tissues becoming brown or black.
- Galls can also appear on roots, stem, branches, petioles and leaf veins.
- In early stages, tumours are more or leass spherical, white colored and quite soft.Later , the outer tissue becomes dark brown.
- Gall tissue is disorganized growth with an enlarged cambium and irregular vascular tissue.
- Movement of water and nutrients is severly impaired by galls.
- Attacked plants may remain stunted producing small chlorotic leaves. Main roots grow poorly and their yiled is reduced (Figure-4).

Etiology

Scientific Classification

Kingdom : Bacteria

Division : Gracilicutes

Class : Proteobacteria

Order : Rhizobiales

Family : Rhizobeaceae

Genus : *Agrobacterium*

Species : *tumefaciens*

- Shape: Rod shaped.
- Staining: Gram negative.
- Size: 0.6-1.0 µm x 1.5-3.0 µm.
- Nature: Aerobic.
- Donot produce endospores.
- Flagella: 2-4, peritrichous flagella.
- Plasmids: Pathogen contain one to several large plasmids composed of circular double stranded DNA.
- Ti-Plasmid: Plasmid carries the genes for gall or tumour induction and is called tumour inducing plasmid (Ti-Plasmid).
- Most important characteristic feature of pathogen is its ability to transfer part of Ti-plasmid (T-DNA) into host plant chromosome and transform normal plant cells to tumor cells.
- This has two principal physiological effects; it disrupt the metabolism of growth hormones in the tissues, brining about oncogenesis and it causes the synthesis of specific chemicals called opines. These opines can be utilized only by a bacteria that contain an appropriate Ti-plasmid and make it a 'genetic parasite'.
- High phytohormone level results from expression of three oncogenes transferred stably into the plant genome from *A. tumefaciens* : iaaM, iaaH and ipt.

- iaaM and iaaH oncogenes direct auxin biosynthesis and ipt oncogene cause cytokinin production.

Disease Cycle

- Pathogen is a soil inhabitant and perennates in infested soil where it can live as a saprophyte for several years.
- When susceptible plants are raised at such sites , the pathogen reaches the crown region through the freshly developed wounds due to cultural practices, grafting or by insects.
- Once bacteria reaches inside the tissue, produce an enzyme that cuts the Ti-plasmid at specific sites releasing the segment of T-DNA form the Ri or Ti plasmid is integrated into a plant cell chromosome at various site and expressed.
- T-DNA expression results in overproduction of plant hormones and usually the synthesis of opines. Uncontrolled synthesis of hormones stimulates plant cells to divide , enlarge and form a tumor.
- With the increase in number and size of tumor cells,tumours exert pressure on surrounding and underlying normal tissues, which may be distorted xylem vessels and ultimately impaired the water supply to upper parts of plants .
- Due to invasion of tumours by secondary microbes decay of peripheral cells takes place.
- The bacterium from these galls may escape to contaminate the surrounding soil of healthy roots.
- The pathogen is then disseminated to new planting sites and healthy plants by splashing rain, irrigation water, tools, wind, insects and plant parts used for propagation.

Epidemiology

- Temperature- 21 to 23^{0}C.
- Warm cloudy and rainy weather.

Management

- Grow resistant varieties.
- Avoid wounding the crown and control root chewing insects.

- Foliar spray with Bordeaux mixture, Streptocylin or Terramycin.
- Regulate soil fertility to reduce succulence of new growth.
- Deep seeds or root systems of nursery seedlings or root stocks upto crown region in *A. tumefaciens* strain K84 suspension.

Model Question Paper

A. Objective Type Questions

a. Choose the correct answer from the following

(1) *Ventruia inaequalis* belongs to

a. Pleosporales b. Peronosporales

c. Melanconiales d. Moniliales

(2) Scab of apple is caused by

a. *Venturia inaequalis* b. *Rhizoctonia bataticola*

c. *Sphacelotheca sorghi* d. *Sphacelotheca cruenta*

(3) Fire blight of apple is caused by

a. *Erwinia amylovora* b. *Uncinula necator*

c. *Sphacelotheca sorghi* d. *Sphacelotheca cruenta*

(4) Powdery mildew of apple is caused by

a. *Podosphaera leucotricha* b. *Oidium mangiferae*

c. *Sphacelotheca sorghi* d. *Sphacelotheca cruenta*

(5) The pathogen *Ventruia inaequalis* produces

a. Oospores b. Cleistothecium

c. Pseudotheicum d. Apothecium

(6) *Ventruia inaequalis* represents the anamorphic state of ascomycetous fungus

a. *Spilocaea pomi* b. *Oidium mangiferae*

c. *Sphacelotheca sorghi* d. *Sphacelotheca cruenta*

(7) The fruiting body cleistothecia formed in which fungus

a. *Podosphaera leucotricha* b. *Puccinia graminis*

c. *Sphacelotheca sorghi* d. *Sphacelotheca cruenta*

(8) White superficial powdery fungal growth appear on leaves, stalks of panicles, flowers and young fruits of grape is characteristic symptoms of

a. Powdery mildew b. Bud rot

c. Grey blight d. None

(9) The pathogen *Podosphaera leucotricha is*

a. Biotroph b. Hemibiotroph

c. Necrotroph d. None

(10) *Podosphaera leucotricha* belongs to order.

a. Erysiphales b. Peronosporales

c. Melanconiales d. Moniliales

(11) White superficial powdery fungal growth appear on leaves, stalks of panicles, flowers and young fruits of apple is characteristic symptoms of

a. Powdery mildew b. Bud rot

c. Grey blight d. None

(12) Which pathogen has the distinction of being the first bacterium of plants

a. *Erwinia amyovora* b. *Xanthomonas oryzae*

c. *Pseudomonas solanacearum* d. *Bacillus subtilis*

(13) The first pathogen to be associated with an inscet vector.

a. *Erwinia amyovora* b. *Xanthomonas oryzae*

c. *Pseudomonas solanacearum* d. *Bacillus subtilis*

(14) The *Erwinia amyovora* can be a plant pathogen was given by

a. T.J.Burrill b. E.F Smith

c. Adolf Mayer d. E J Butler

(15) Small tumours or galls develop in the wound just below the soil surface in apple is characteristic symptoms of

a. Powdery mildew b. Bud rot

c. Grey blight d. Crown gall

(16) Plasmid carries the genes for gall or tumour induction and is called

a. Ti-Plasmid
b. Gi-Plasmid
c. RI-Plasmid
d. PI-Plasmi

Answer

S. No	Answer	S. No	Answer
1	a. Pleosporales	9	a. Biotroph
2	a. *Venturia inaequalis*	10	a. Erysiphales
3	a. *Erwinia amylovora*	11	a. Powdery mildew
4	*a. Podosphaera leucotricha*	12	a. *Erwinia amyovora*
5	c. Pseudotheicum	13	a. *Erwinia amyovora*
6	a. *Spilocaea pomi*	14	a. T.J.Burrill
7	a. *Podosphaera leucotricha*	15	d. Crown gall
8	a. Powdery mildew	16	a. Ti-Plasmid

b. Fill in the blanks with suitable word (s)

1. Apple scab is caused by _______________ is telomorph, whereas, anamorph of the fungus is _______________.
2. Apple scab disease perinnates in _____________and primary infection is caused by _______________.
3. Leaf ___________period and ______________plays important role in primary infection of apple scab.
4. Secondary source of inoculum for scab development is ___________.
5. Application of __________to leaf litter during fall (when more than 80% leaves have fallen) hasten decomposition of leaves due to microbial action and reduces primary inoculum.
6. In case of fire blight of apple, tips of young infected shoots wilt, forming a very typical _________________symptom.
7. The fire blight bacteria _____________in bark tissues along the edges of cankers caused by infections during previous years.
8. The *Erwinina amylovora* penetrate the tree through ___________or _______________.
9. Secondary infection of *Erwinina amylovora* arises from ________of fresh infections.
10. *Podosphaera leucotricha* overwinters in the _____________that had been infected in the preceding growing season.

Answer

Sl. No	Answer	Sl. No	Answer
1	*Venturia inaequalis* (telomorph), *Spilocaea pomi* (Anamorph)	6	Shepherd's crook
2	Pseudothecia, Ascospores	7	Overwinter
3	Wetness, temperature	8	Natural openings or wounds
4	Airborne conidia	9	Ooze
5	5% urea	10	Dormant bud

c. State whether the following statements are True or False

1. The imperfect stage (*Spilocaea pomi*) survive on living apple plant parts and perfect (*Ventruia inaequalis*) on fallen dead leaves.
2. *Venturia inaequalis* ia an obligate parasite.
3. *Erwinia amylovora* is a Gram +ve bacteria.
4. The fungus *Venturia inaequalis* is bipolar heterothallic.
5. The ascocarp of pathogen*Oidium farinosum* is known as cleistotheica.
6. White floury patches symptom appear on both side of the leaves of grape in powdery mildew disease.
7. *Erwinia amylovora* pv.*pyri* has the distinction of being the first bacterium of plants shown to be a pathogen.
8. Plasmid carries the genes for gall or tumour induction and is called tumour inducing plasmid (Ti-Plasmid).
9. *Agrobacterium tumefaciens* as an endospore forming bacteria.
10. The telemorphic stage of *Spilocaea pomi is Venturia inaequalis.*

Descriptive Question

(a) Long answer questions

1. Explain in detail the occurrence, importance, symptoms, etiology, disease cycle and management of scab of apple.
2. Briefly describe the diagnostic symptoms, epidemiology & management of any two of the following grape diseases :

   (i) Fire blight (ii) Crown gall

3. Describe the powdery mildew of apple in the following headings:

   (i) Pathogen (ii) Symptoms

   (iii) Disease cycle (iv) Disease management

4. Describe in detail the most distinguishing symptoms, causal organism, disease cycle and management of crown gall of apple.

5. Give the diagrammatic representation of the disease cycle of the following diseases of apple

   (i) Powdery mildew (ii) Scab

   (iii) Fire blight

6. Name two bacterial diseases of apple with their causal organism. Describe etiology, disease cycle and management of any one disease.

(b) Short answer questions

1. Write down the etiology of powdery mildew of apple.
2. Which fungal pathogen causes crown gall of apple? Give its systematic position.
3. Write down the management practices of scab of apple.
4. Describe the causal organisms and mode of penetration of firelight of apple.
5. Why the downy mildew disease of grape is historically important ?
6. Suggest some chemicals for management of scab of apple.
7. Which type of disease can be effectively managed by mixed cropping.

Fig. 1: Scab

Fig. 2: Powdery mildew

Fig. 3: Fire blight

Fig. 4: Crown gall

Plate-20: Photograph showing symptoms of major diseases of apple

21

Peach Crop Diseases & Management

Name of Diseases	Causal Organism
Leaf curl	*Taphrina deformans (Berk)* Tulansne
Leaf spot	*Phyllosticta cerasicola* Speg.
Powdery mildew	*Sphaerotheca pannosa* (Wallr.) Lév, *Podosphaera leucotricha* (Ellis &Everh.) E.S. Salmon
Whiskar rot	*Rhizopus stolonifer* Ehrenb

1) Leaf Curl

Taphrina deformans

Disease Symptoms

- Initially yellowish to reddish areas appear on leaf blade of young leaves in the spring.
- These areas progressively thicken and pucker along the midrib causing the leaf to curl.
- Infected leaves often develop red to purple color. Later on white to silvery coating of spores develop on the upper leaf surface and the affected leaves become crisp and brittle.
- Finally, infected leaves abscise prematurely or may sometimes remain attached , gradually turning dark brown on severly infected trees.
- Fruit infection is characterized by irregular, raised, wrinkled and reddish lesions.
- Infected flowers and fruits drop rapidally(Fig-1).

Etiology

Scientific Classification

Kingdom : Fungi

Division : Ascomycota

Class : Archiarcomycetes

Order : Taphrinales

Family : Protomycetaceae

Genus : *Taphrina*

Species : *deformans*

- Mycelium: septate with typically binucleate cells, intercellular, mostly subcuticular,sub epidermal or may proceed deeper into the tissues.After maturity, the hyphal cells break up , become free form the neighbouring ones and are called chlaymydospores or ascogenous cells.
- Asci : naked clynideric, clavate, rounded or truncate at the apex and have a stalk cel size, measuring 17- 56x5-7μm and form compact and white hymenium.
- Ascus: contains eight ascopores which are round, ovate, elliptic and measure 3-7 μm in diameter.
- Ascospores: multiply by budding inside of outside the ascus, producing condia.
- Blastospores: (bud conidia) formed by budding from ascospores are oval, minute and measure 2.5 x4.8 μm, may germinate to produce mycelium.

Disease Cycle

- Pathogen survives as blastospores in the crevices of the bark and under the bud scales. of the tree.
- Priamry infection occur in spring so called as ‘spring time’ when the bud swell and the first leaf emerges from it.
- Blastospores germinate and infect the yound leaves.
- Hyphae that grow intercellularly stimulate the adjacent cells to form the the neoplastic growth, rsulting in leaf curl symptoms.
- The hyphae later aggregate below the cuticle and breakup to form ascogenous cells, which develops into an ascus.
- Ascospores are released forcibly, multiply by budding inside and outside the ascus forming blastospores.
- These collect on the leaf surface giving rise it a powdery and wrinkled appearance.
- There is no secondary infection.

Epidemiology

- Relative humidity- >85%.
- Optimum Temperature – 15- 20^0 C.

Management

- Select resistant varieties like Stark. Early Giant, StarkingFelicious, TasisSemisto, World Earliest, Shan-e-Punjab, Bed Wills early, July Alberta whenever possible.
- Foliar spray with Sulfur @ 0.2% or Mancozeb @ 0.2%,Blue copper or Blitox 50 @ 0.3percent . For best results, trees should be sprayed to the point of runoff or until they start dripping.
- Keep the ground beneath the trees raked up and clean, especially during winter months.
- Prune and destroy infected plant parts as they appear.
- If disease problems are severe, maintain tree health and vigor by cutting back more fruit than normal, watering regularly (avoiding wetting the leaves if possible) and apply an organic fertilizers high in nitrogen.

Model Question Paper

A. Objective Type Questions

a. Choose the correct answer from the following

(1) *Taphrina deformans* belongs to order

a. Exobasidiales b. Peronosporales
c. Taphrinales d. Moniliales

(2) Leaf curl is caused by

a. *Mycosphaerella fragariae* b. *Rhizoctonia bataticola*
c. *Exobasidium vexans* d. *Sphacelotheca cruenta*

(3) How many ascospores are present in asci of *Taphrina deformans*

a. 8 b. 6
c. 2 d. 4

(4) The spore formed by budding from ascospores of *Taphrina deformans* are called

a. Blastospores
b. Uredospore
c. Teliospore
d. Aeciospore

(5) Optimal epidemiological condition for development of leaf spot of peach is

a. Relative humidity- >85%, Optimum temperature – 15- 20^{0}C
b. Relative humidity- >75%, Optimum temperature – 25- 30^{0}C
c. Relative humidity- >65%, Optimum temperature – 30- 40^{0}C
d. Relative humidity- >55%, Optimum temperature – 10- 15^{0}C

Answer

S. No	Answer	S. No	Answer
1	a.Taphrinales	4	a.Blastospores
2	*Mycosphaerella fragariae*	5	a. Relative humidity- >85%, Optimum Temperature – 15- 20^{0}C
3	a. 8	6	---

b. State whether the following statements are True or False

1. The hyphal cells of *Taphrina deformans* break up in peach crop, become free form the neighbouring ones are called chlaymydospores.
2. The asci of *Taphrina deformans* are naked clynideric, clavate, rounded or truncate at the apex form compact and white hymenium.
3. Blastospores formed by budding from ascospores of *Taphrina deformans* are oval, minute, may germinate to produce mycelium.
4. *Taphrina deformans* survives as blastospores on various parts of the peach.
5. *Taphrina deformans* belongs to class Archiarcomycetes.
6. The ascus of *Taphrina deformans* contains eight ascopores.
7. There is no secondary infection in leaf curl disease.
8. Ascospores of *Taphrina deformans*are released forcibly, multiply by budding inside and outside the ascus forming blastospores.

Answer

S. No	Answer	S. No	Answer
1	True	5	True
2	True	6	True
3	True	7	True
4	True	8	True

B. Descriptive Questions

a. Long answer questions

1. Describe in detail the most distinguishing symptoms, causal organism, disease cycle , epidemiology and management of Leaf curl of peach.
2. Explain in detail about symptoms, etiology, disease cycle and management of Leaf curl of peach.
3. Describe the Leaf curl of peach in the following headings:

   (i) Pathogen (ii) Symptoms

   (iii)Disease cycle (iv) Disease management

b. Short answer questions

1. Which fungal pathogen causes Leaf curl of peach.? Give its systematic position.
2. Write down the management practices of Leaf curl of peach.
3. Describe the causal organisms and diagnostic symptoms of Leaf curl of peach.
4. Write down the etiology of Leaf curl of peach.

Fig. 1: Leaf curl

Plate- 21: Photograph showing symptoms of major diseases of peach

22

Grape Crop Diseases & Management

Name of Disease	Causal organism
Downy mildew	*Plasmopara viticola*
Powdery mildew	*Uncinula necator*
Anthracnose	*Gloeosporium ampelophagum* (*Elsinoe ampelina)*

1. Downy mildew

Diagnostic Symptoms

- Symptoms appear on all aerial and tender parts of the vine.
- Irregular, yellowish, translucent spots on the upper surface of the leaves.
- Correspondingly on the lower surface, dirty white, powdery growth of fungus appears.
- Affected leaves become, yellow and brown and gets dried due to necrosis
- Premature defoliation.
- Dwarfing of tender shoots.
- Infected leaves, shoots and tendrils are covered by whitish growth of the fungus.
- White growth of fungus on berries which subsequently becomes leathery and shrivels. Infected berries turn hard, bluish green and then brown.
- Later infection of berries results in soft rot symptoms. Normally, the fully grown or maturing berries do not contact fresh infection as stomata turn non-functional.
- No cracking of the skin of the berries (Fig.1). .

Etiology

Scientific Classification

Kingdom : Fungi

Division : Eumycota

Class : Oomycetes

Order : Peronosporales

Family : Peronosporaceae

Genus : *Plasmopara*

Species : *viticola*

- Pathogen:*Plasmopara viticola.*
- The pathogen is biotroph.
- Mycelium is intercellular with spherical haustoria, coenocytic, thin walled and hyaline.
- Sporangiophores arise from hyphae in the sub stomatal spaces.
- It branched at right angle to the main axis and at regular intervals. Secondary branches arise from lower branches.
- The sporangia are thin walled, oval or lemon shaped.
- The sporangia germinate either by forming zoospores, or behave as conidia and germinate by germ tube.
- The Zoospores are egg shaped, biflagellate and 7 – 9 micron meter. The oospores are thick walled.
- The oospores are thick walled with a rough epispore and are produced mostly in tissues adjacent to the midrib.

Disease Cycle

- Pathogen overwinters mainly as oospores in the debris in the soil, serve as the primary inoculum.
- They germinate and form numerous biflagellate zoospores.
- Primary infection is by zoospores transported by wind or water to the wet lower leaves near the soil line.

- They encyst and form a germ tube that penetrates through stomata or the epidermis.
- Infection becomes systemic and symptom appear after 10-15 days of infection.
- Secondary infection takes place by the motile zoospores by splashing rain.
- As the close of the season, oospores are formed, which fainally reach the soil.

Epidemiology

- Optimum temperature : 20-22°C.
- Relative humidity : 80-100 per cent.
- Wet winter followed by a wet spring and a warm summer with intermittent rains.

Management

- Deep ploughing of fields during summer.
- Make sure soils are well drained.
- Field sanitation.
- Remove weed and plant residues from the field.
- Keep tools and equipment clean.
- Apply a balanced fertilizer.
- Destroy the alternate host plants as well as crop debris .
- Vine should be kept high above ground to allow circulation of air by proper spacing.
- Pruning (April- May & September –October) and burning of infected twigs.
- Grow resistant varieties like Amber, Queen, Cardinal, Champa, Champion, Dogridge and Red Sultana.
- Foliar spray with Copper oxychloride 50% WG@ 0.24% or Cymoxanil 50% WP@ 0.24% or Mancozeb 75% WP @ 0.2% or Propineb 70% WP @ 0.30% or Cymoxanil 8% + Mancozeb 64% WP @ 0.2% or Metalaxyl M 4%+ Mancozeb 64% WP @ 0.25% or Metalaxyl 8% + Mancozeb 64% WP@ 0.2%,if required repeat after 15 days.

2. Powdery Mildew : *Uncinula necator*

Diagnostic Symptoms

- Powdery growth mostly on the upper surface of the leaves.
- Malformation and discolouration of affected leaves.
- Discolouration of stem to dark brown.
- Floral infection results in shedding of flowers and poor fruit set.
- Early berry infection results in shedding of affected berries.
- Powdery growth is visible on older berries and the infection results in the Cracking of skin of the berries (Fig.2).

Etiology

Scientific Classification

Kingdom : Fungi

Division : Eumycota

Class : Leotiomycetes

Order : Erysiphales

Family : Erysiphaceae

Genus : *Uncinula*

Species : necator

- The fungus is oidium type.
- White growth consists of mycelium, conidiophores and conidia.
- Mycelium is external, septate and hyaline.
- Conidiophores are short and arise from external mycelium.
- Conidia are produced in chain. They are single celled, hyaline and barrel shaped.
- When the mating types are present cleistothecia can form on all infected tissues during later part of the growing season.
- Mature cleistothecia are black, depressed, contains 4-8 asci.
- 4-6 oval ascospores occur within each ascus.

Disease Cycle

- Pathogen overwinters as mycelium in the dormant buds and as cleistotheica in the cracks in the bark.Besides, the collateral hosts and volunteer plants provide conidia as the primary mycelium.
- Infected buds give rise to systematically infected shoots.
- Cleistothecia, when wet, discharge the ascospores, which infect the leaves.
- Ascospores and conidia germinate and push haustoria into the epidermal cells from the appresorium.
- Mycelium grows externally and forms conidia in 7 to 10 days after infection.
- Secondary infections can occur through spores produced, released when conditions are favorable for growth of the fungus in spring and cause new infections.

Favourable Conditions

- Cool dry weather.
- Optimum temperature – 20-25 ^{0}C.
- Enough relative humidity.

Management

- Deep ploughing of fields during summer .
- Use tolerant varieties if available.
- Keep a wide distance between vines to allow for good air circulation.
- Alternativley chose pruning practices that favor an open canopy.
- Monitor the field regularly for symptoms of the disease.
- Use nitrogen fertilizers with caution to avoid excessive vegetative growth.
- Field sanitation.
- Destroy the alternate host plants as well as crop debris.
- Adopt ecological engineering by growing the attractant, repellent, and trap crops around the field bunds.

- Grow resistant varieties like ChholthRed ,Skibba Red, Chholth white.
- Foliar spray with Wettable Sulphur @0.3% or Karathane or Calixin @ 0.1% orCarbendazim 46.27% SC@ 0.1% or Flusilazole 40% EC@ 0.01% or Myclobutanil 10% WP @ 0.04% or Sulphur 55.16 % SC @ 0.30% or Triadimefon 25% WP@ 0.010% , sprayed at 15 days interval .

3. Anthracnose

Diagnostic Symptoms

- Symptoms visible on leaves, stem, tendrils and berries.
- Young shoots and fruits are more susceptible than leaves.
- Circular, greyish black spots or red spots with yellow halo appear.
- Later the centre of the spot becomes grey, sunken and fall off resulting in a symptom called 'shot hole'.
- Black, sunken lesions appear on young shoots.
- Cankerous lesions on older shoots. Girdling and death of shoots occur.
- Infection on the stalk of bunches and berries result in the shedding of bunches and berries respectively.
- Sunken spots with ashy grey centre and dark margin on fruits (Birds eye symptom).
- Mummification and shedding of berries (Fig.3).

Etiology

Scientific Classification

Kingdom : Fungi

Division : Eumycota

Phyllum : Deuteromycotina

Class : Coelomycetes

Order : Melancoliales

Family : Melanconiaceae

Genus : *Gloeosporium*

Species : *ampelophagum*

- Pathogen: *Gloeosporium ampelophagum* (*Elsinoe ampelina).*
- Perfect stage : *Elsinoe alpelina.*
- Hyphae: well developed, septate and branched.They accumulate in subcuticular region of infected plant and develop acervuli.
- Acervuli: single celled, colorless, delicate and small conidiophores are formed in large number in each acervulus.
- Conidia: produce on tips of conidiophores, unicellular, hyaline, slightly narrower in the middle than at the ends, measure 5-6 x2.5-3.5μm.

Disease Cycle

- Pathogen survives in infected plant debris and act as a source of inoculums.
- Primary infection is through spores produced by disease vines.
- Secondary infection by wind borne conidia.

Favourable Conditions

- Optimum temperature-24-26^{0}C.
- Leaf wetness period-3 hrs.
- Humid and wet weather.

Management

- Choose sites with suitable sun exposure and good air circulation.
- Removal of infected twigs
- Use of resistant varieties such as– Bangalore Blue, Beauty Seedless, Bharat Early, Golden queen, Large white.
- Make sure to keep a wide distance between vines.
- Remove any wild grapes near the vineyard.
- Monitor the vines and remove fruit or plant parts that show sign of the disease.
- Prune the vines in early winter during dormancy.
- Remove plant residues from the vineyard plow field and burry.
- Foliar spray with Aureofungin 46.15%w/v. SP@ 0.005% or Carbendazim

50%WP@ 0.1% or Chlorothalonil 75% WP@ 0.2% or Kitazin 48% EC @ 0.20% or Mancozeb75% WP @ 0.25%.

Model Questions

A. Objective Type Questions

a. Choose the correct answer from the following

(1) *Colletotrichum gloeosporioides* belongs to oomycete.

a. Peronosporales b. Albuginales

c. Melanconiales d. Moniliales

(2) Downy mildew of grape is caused by

a. *Plasmopara viticola* b. *Rhizoctonia bataticola*

c. *Sphacelotheca sorghi* d. *Sphacelotheca cruenta*

(3) Anthracnose of grape is caused by

a. *Gloeosporium ampelophagum* b. *Uncinula necator*

c. *Sphacelotheca sorghi* d. *Sphacelotheca cruenta*

(4) Powdery mildew of grape is caused by

a. *Uncinula necator* b. *Oidium mangiferae*

c. *Sphacelotheca sorghi* d. *Sphacelotheca cruenta*

(5) Acervuli of *Colletotrichum gloeosporioides* may posses sterile hair like structures called

a. Oospores b. Cleistothecium

c. Setae d. Apothecium

(6) *Gloeosporium ampelophagum*represents the anamorphic state of ascomycetous fungus

a. *Elsinoe alpelina* b. *Oidium mangiferae*

c. *Sphacelotheca sorghi* d. *Sphacelotheca cruenta*

(7) Hyphae of *Colletotrichum gloeosporioides*well developed, septate and branched, subcuticular region of infected plant and develop

a. Acervuli b. Cleistothecium

c. Setae d. Apothecium

(8) Optimum epidemiological condition for development of powdery mildew of grape

a. Optimum temperature- 15-30^{0}C, Relative humidity-60-90 %.

b. Optimum temperature- 15-40^{0}C, Relative humidity-50-70 %.

c. Optimum temperature- 15-50^{0}C, Relative humidity-40-60 %.

d. Optimum temperature- 15-20^{0}C, Relative humidity-40-50 %.

(9) White superficial powdery fungal growth appear on leaves, stalks of panicles, flowers and young fruits of grape is characteristic symptoms of

a. Powdery mildew b. Bud rot

c. Grey blight d. None

(10) The pathogen *Plasmopara viticola is*

a. Biotroph b. Hemibiotroph

c. Necrotroph d. None

(11) Erysiphales belongs to order.

a. Erysiphales b. Peronosporales

c. Melanconiales d. Moniliales

(12) White superficial powdery fungal growth appear on leaves, stalks of panicles, flowers and young fruits of grape is characteristic symptoms of

a. Powdery mildew b. Bud rot

c. Grey blight d. None

(13) Birds eye symptoms found on grape fruit infected with which disease

a. Powdery mildew b. Bud rot

c. Downy mildew d. Anthracnose

Answer

S. No	Answer	S. No	Answer
1	c. Melanconiales	7	a. Acervuli
2	a. *Plasmopara viticola*	8	a. Optimum temperature- 15-30^{0}C, Relative humidity-60-90 %.
3	a. *Gloeosporium ampelophagum*	9	a. Powdery mildew
4	a. *Uncinula necator*	10	a. Biotroph --
5	c. Setae	11	a. Erysiphales
6	a. *Elsinoe alpelina*	12	a. Powdery mildew
		13	d. Anthracnose

(b) State whether the following statements are *True* or *False*

1. Secondary infection of anthracnose is mainly due to transfer of conidia through rain splash or wind driven rain water.
2. Hyphae of *Colletotrichum gloeosporioides* accumulate below the host cuticle and develop fruiting bodies called acervuli.
3. Fruits infected with Anthrcnose at mature stage carry the pathogen into storage and cause considerable loss during storage, transit and marketing.
4. Secondary spread of powdery mildew disease can occur thorugh air borne conidia produced in these infections.
5. White growth of *Plasmopara viticola* on berries of grapes which subsequently becomes leathery and shrivels.
6. Foliar spray of 0.02% Streptomycin followed by Copper oxychloride (0.3%) manages bacterial blight of mango.
7. Foliar sprays with wettable sulphur@ 0.3% control powdery mildew of grape.
8. The sporangia of *Plasmopara viticola* are thin walled, oval and lemon shaped.
9. The zoospores of *Plasmopara viticola* are egg shaped and biflagellate.
10. The oospores of *Plasmopara viticola* are thick walled with a rough epispore and are produced mostly in tissues adjacent to the midrib of leaves of grape.

Answer.

S. No	Answer	S. No	Answer
1	True	6	True
2	True	7	True
3	True	8	True
4	True	9	True
5	True	10	True

B. Descriptive questions

a. Long answer questions

1. Describe in detail about symptoms, etiology, disease cycle and management of powdery mildew of grape.
2. Describe in detail the most distinguishing symptoms, causal organism, disease cycle and management of anthracnose of grape.
3. Describe in detail the most distinguishing symptoms, causal organism, epidemiology, disease cycle and management of downy mildew of grape.
4. Briefly describe the diagnostic symptom and management of powdery mildew and anthracnose of grape.
5. Explain the downy mildew of grape in the following headings:

   (i) Pathogen (ii) Symptoms

   (ii) Disease cycle (iv) Disease management

b. Short answer questions

1. Which fungal pathogen cause anthracnose disease of grape? Give its systematic position.
2. Elaborate disease cycle of Powdery mildew of grape in pictorial form .
3. Describe in brief about managment of downy mildew of grape.
4. How would you identify the powdery mildew of grape.
5. Write the epidemiological condition for development of powdery mildew of grape.
6. Write down the systemic position and etiology of pathogen *Colletotrichum gloeosporioides* causingAnthracnose of grape.

c. Very short answer questions

1. Discuss about epidemiology of powdery mildew of grape.
2. Differentiate between downy mildew and powdery mildew.
3. Name the three fungal diseases of mango with their causal organism.
4. Name the pathogen or causal organism of following disease ;
   (i) Downy mildew (ii) Powdery mildew
5. Write short notes on management practices of Powdery mildew.

Fig.1. Downy mildew

Fig. 2: Powdery mildew

Fig. 3: Anthracnose

Plate-22: Photograph showing symptoms of major diseases of grape

23

Strawberry Crop Diseases & Management

Name of Diseases	Causal Organism
Leaf spot/Common leaf spot	*Mycosphaerella fragariae*(Tul.) Lindau
Powdery mildew	*Podosphaera aphanis*
Vertillium wilt	Verticillium spp.
Botrytis blight	*Botrytis cinerea*

1) Leaf Spot

(*Mycosphaerella fragariae* teleomorph, *Ramularia brunnea* Peck anamorph).

Disease Symptoms

- Older leaves usually show purple spots (3-6 mm in diameter) on the upper surface, sometimes with a slightly darker halo.
- As the spots mature, they become white to gray and surrounded by a brownish halo.
- A typical lesions, uniformly brown without darker border or lighter centers, may form on young leaves.
- Later on, the whole leaf is covered by numerous lesions and become chlorotic, withered and die.
- Fruits are generally not directly infected, but the loss of foliage can reduce their quality and yield (Fig-1).

Etiology

Scientific Classification

Kingdom : Fungi

Division : Ascomycota

Class : Dothideomycetes

Order : Capnodiales

Family : Mycospharellaceae

Genus : *Mycosphaerella*

Species : *fragariae*

- Teleomorph of *Mycosphaerella fragariae* reproduces by forming perithecia that are black, partially embedded, globose (100-150 μm) and minutely ostiolate.
- Asci: Size 30-40 x 10-15 μm, cylindrical to clavate, have short stalks, and contain 8 spores, are formed in small clusters.
- Ascospores: Size12-14 x 3-4 μm, hyaline and two-celled with a median septum with each cell generally containing two oil drops .
- Anamorph of *M. fragariae, Ramularia brunnea* produces conidia that are elliptical to cylindrical (20-40 x 3-5 μm) hyaline, and zero- to four-septate.
- Conidia are often formed in short chains and are borne on short, hyaline, unbranched, and frequently curved conidiophores .
- Hyphae of *M. fragariae* may aggregate and form sclerotia which resemble perithecia in size and shape but lack a cavity.

Disease Cycle

- Pathogen overwinter on infected leaf debris on the soil.
- Resting spores produces conidia which are the source of primary infection.
- Secondary spread occurs throughspores that are carried to new leaves by rain splashes and wing.

Epidemiology

- Relative humidity- >85%.
- Optimum temperature – Around 25^{0} C.
- Prologned leaf wetness.

Integrated Management

- Use seeds for planting from certified sources.
- Plant resistant varieties, if available in area.

- Plant in well drained light soil with good air circulation.
- Clear the fields and surroundings of weeds.
- Apply a balanced fertilizer program without excess nitrogen.
- Foliar spray with Chlorothalonil 75% WP @ 0.2% or Copper oxychloride 50% WP @ 0.3% or Mancozeb 75% WP@ 0.2% or Hexaconazole 2% SC @ 0.1% as required depending upon crop stage (second spray after 15 days interval) .
- Do not water in the evening to avoid high humidity conditions.
- Remove infected plants and crop debris and burn or bury them at some distances of the field.
- Do not work in field when foliage is wet and avoid injuries the plants.

Model Question Paper

A. Objective Type Questions

a. Choose the correct answer from the following

(1) *Mycosphaerella fragariae*belongs to order.

a. Exobasidiales b. Peronosporales

c. Capnodiales d. Moniliales

(2) Leaf spot of strawberry is caused by

a. *Mycosphaerella fragariae* b. *Rhizoctonia bataticola*

c. *Exobasidium vexans* d. *Sphacelotheca cruenta*

(3) Leaf spot of strawberry is a type of disease

a. Monocyclic b. Polycyclic

c. Polyetic d. None

(4) *Mycosphaerella fragariae* reproduces by forming:

a. Perithecia b. Cleistothecia

c. Apothecia d. Chlamydospore

(5) Leaf spot of strawberry diseases is

a. Soil borne b. Seed borne

c. Air borne d. All

(6) The anamorphic stage of *Mycosphaerella fragariae is*

a. *Ramularia brunnea* b. *Rhizoctonia bataticola*

c. *Exobasidium vexans* d. *Sphacelotheca cruenta*

(7) The teleomorphic stage of *Ramularia brunnea is*

a. *Mycosphaerella fragariae* b. *Rhizoctonia bataticola*

c. *Exobasidium vexans* d. *Sphacelotheca cruenta*

(8) Hyphae of *Mycosphaerella fragariae* may aggregate and form:

a. Perithecia b. Cleistothecia

c. Apothecia d. Sclerotia

Answer

S. No	Answer	S. No	Answer
1	c.Capnodiales	5	a. Soil borne
2	a. *Mycosphaerella fragariae*	6	a. *Ramularia brunnea*
3	b.Polycyclic	7	a. *Mycosphaerella fragariae*
4	a.Perithecia	8	d. Sclerotia

b. State whether the following statements are True or False.

1. Pathogen *Mycosphaerella fragariae*overwinter on infected leaf debris in the soil.
2. Resting spores *Mycosphaerella fragariae* produces conidia which are the source of primary infection.
3. Secondary spread leaf spot of strawberry occurs through spores that are carried to new leaves by rain splashes and wind.
4. Teleomorph *Mycosphaerella fragariae* reproduces by forming perithecia that are black, partially embedded, globose and minutely ostiolate.
5. The asci of pathogen *Mycosphaerella fragariae* are cylindrical to clavate, have short stalks, and contain 8 spores.
6. *Mycosphaerella fragariae*anamorph conidia from *Ramularia brunnea* are elliptical to cylindrical, 20–40 x 3-5 µm, hyaline, and zero–four septate.
7. The ideal temperature for the development of strawberry leaf spots is around 25^0 C and the relative humidity should be greater than 85%.

8. Hyphae of *Mycosphaerella fragariae* may aggregate and form sclerotia which resemble perithecia in size and shape but lack a cavity.

Answer

S. No	Answer	S. No	Answer
1	True	6	True
2	True	7	True
3	True	8	True
4	True	9	True

B. Descriptive Questions

a. Long answer questions

4. Describe in detail the most distinguishing symptoms, causal organism, disease cycle , epidemiology and management of leaf spot of strawberry.
5. Explain in detail about symptoms, etiology, disease cycle and management of leaf spot of strawberry.
6. Describe the leaf spot of strawberry in the following headings:

   (i) Pathogen (ii) Symptoms

   (iii)Disease cycle (iv) Disease management

b. Short answer questions

1. Which fungal pathogen causes leaf spot of strawberry.? Give its systematic position.
2. Write down the management practices of leaf spot of strawberry.
3. Describe the causal organisms and diagnostic symptoms of leaf spot of strawberry.
4. Why the leaf spot of strawberry is historically important in India?
5. Write down the etiology of pathogen causing leaf spot of strawberry.

Fig. 1: Leaf spot

Plate-23: Photograph showing symptoms of major diseases of strawberry

24

Coconut Crop Diseases & Management

Sl.No	Name of Diseases	Causal Organism
1	Bud rot	*Phytophthora palmivora*
2	Ganoderma Wilt	*Ganoderma lucidum*
3	Stem Bleeding	*Ceratocystis paradoxa/ Ophiostoma paradoxa*
4	Grey Blight	*Pestalotiopsis palmarum*

1) Bud Rot

Diagnostic Symptoms

- Severe on young palms.
- Yellowish green discoloration of the heart leaf or crown leaf.
- The basal tissues of the leaf rot quickly and can be easily separated from the crown.
- Spindle withers and droop down.
- Older leaves develop irregular, water soaked spots which are sunken in nature.
- The leaves and sheath in the central spindle fall off leaving an outer whorl of green leaves.
- The withered central shoot can be pulled out very easily from the crown.
- The central crown may rot and in few months the tree may wilt.
- Young nuts fail to mature and fall (Fig.1).

Etiology

- Pathogen: *Phytophthora palmivora.*
- The fungus produces intercellur, non septate, hyaline mycelium.
- Sporangiophores are hyaline and simple or branched occasionally.
- The sporangiophores are hyaline, thin walled, pear shaped with a prominent papillae.

- Sporangia releases reniform, biflagellate zoospores upon germination.
- The fungus also produces thick walled, spherical oospores.
- In addition, thick walled, yellowish brown chlamydospores are also produced.

Disease Cycle

- Primary Infection; Through dormant mycelium, oospores or chlamydospores carried over summer months on the host debris. With the onset of monsoon rains, the fungus becomes active producing cottony mycelium that infects tender host tissue.
- Secondary Infection Through sporangia with numerous zoospores which spread rapidly in the rain water or sporangia spread through wind and insects also.

Epidemiology

- High rainfall, high atmospheric humidity (above 90 per cent), low temperature (18-20°C) and wounds caused by tappper and Rhinoceros beetles.

Management

- Cutting and burning of badly infected palms.
- If the disease is detected early remove the infected portions and protect with Bordeaux paste (Tree surgery).
- Spray copper fungicides (B.M@1% or COC@0.3%) after on set of monsoon toprevent infection
- Give prophylactic spray with 1% Bordeaux mixture to all the healthy plams in the vicinity of diseases one and also before onset of monsoon rains.

2) Ganoderma Wilt

Diagnostic Symptoms

- Called as Basal stem rot /Thanjavur wilt / Bole rot.
- The most usual symptoms are yellowing, withering and drooping of the outer fronds which remain hanging around the trunk for several months before shedding.

- The younger leaves remain green for sometime and later turn yellowish brown.
- The new fronds produced become successively smaller and yellowish in colour which do not unfold properly.
- The wilting plants also show bleeding patches near near the base of the trunk.
- A brown gummy liquid oozes out from the cracks in the tree which slowly result in the death of outer tissues.
- As the infection advances, fresh bleeding patches appear above the old once, up to 3-5 meters height. The decay of the basal portion occurs slowly and tree succumbs to the diseases in 2-3 years (Fig. 2).

Etiology

- Pathogen: *Ganoderma lucidum.*
- The fungus produces a semi circular basidiocarp (bracket), which is attached to the tree with a stalk.
- The bracket is very big about 10-12 cm diameter and woody.
- The upper surface is tough, shining, light to dark brown or almost black with concentric furrows.
- The lower surface is white and soft with numerous minute pores.
- These pores represent the opening of the hymenial tubes, which are lined with basidia and basidio-spores.
- Basidiospores are oval, brown and thick walled.

Disease Cycle

- The fungus is soil-borne and survives in the soil for long time.
- The primary infection is through basidiospores in the soil, which attack roots.
- The irrigation water and rain water also help in the spread of the fungus.

Epidemiology

- Trees grown in sandy loam and sandy soils, water logging during severe rains, low soil moisture content during summer months and damages caused by weevils and beetles.

Management

- Remove and burn severely infected trees which are beyond recovery.
- Isolate the diseased trees by digging a trench all around to check further spread. Irrigate the palms at least once in a fortnight during summer months.
- Apply heavy doses of farm yard manure or compost, green manure at 50 kg/tree/year along with 5 kg of neem cake.
- Drench the soil near the tree with 40 litres of 1 per cent Bordeaux mixture at quarterly interval for thrice a year and repeat after 2-3 years.
- Apply Aureofunginsol 2g+Copper sulphate 1g in100 ml of water or Tridemorph 2ml/100 ml of water through stem injection or root feeding at quarterly intervals.

Model Question Paper

Objective Type Questions

A. Choose the correct answer from the following

(1) *Phytophthora palmivora* belongs to order.

a. Agonomycetales b. Albuginale

c. Peronosporales d. Moniliales

(2) Bud rot of coconut is caused by

a. *Phytophthora palmivora* b. *Rhizoctonia bataticola*

c. *Sphacelotheca sorghi* d. *Sphacelotheca cruenta*

(3) Ganoderma wilt of coconut is caused by :

a. *Cercospora arachidicola* b. *Ganoderma lucidum*

c. *Sphacelotheca sorghi* d. *Sphacelotheca cruenta*

(4) *Phytophthora palmivora* produces thick walled sexual spores

a. Oospores. b. Cleistothecium

c. Pycnidium d. Apothecium

(5) Secondary infection of bud rot of coconut is by.

a. Soil borne conidia b. Air borne sporangia

c. Externally seed borne conidia d. Enternally seed borne conidia

(6) Optimum epidemiological condition for development of Bud rot of coconut

a. Low temperature – 18-20^{0}C, High relative humidity- above 90 percent

b Low temperature – 15-18^{0}C, High relative humidity- below 90 percent

c. Low temperature – 10-12^{0}C, High relative humidity- above 70 percent

d. Low temperature – 8-10^{0}C, High relative humidity- below 70 percent

(7) Coconut plants show bleeding patches near the base of the trunk is a characteristic symptoms of

a. Thanjavur wilt
b. Bud rot
c. Grey blight
d. None

(8) Primary infection of Thanjavur wiltis through which spores in the soil, which attack roots

a. Basidiospores
b. Ascospores
c. Oospores
d. Teliospores

Answer

S. No	Answer	S. No	Answer
1	c. Peronosporales	6	a. Low temperature – 18-20^0 C, High relative humidity- above 90 percent
2	a. *Phytophthora palmivora*	7	a. Thanjavur wilt
3	b. *Ganoderma lucidum*	8	a. Basidiospores
4	a. Acervuli	9	--
5	b. Air borne sporangia	10	--

b. State whether the following statements are *True* or *False*

1. Central shoot of comes off easily on slight pulling in coconut plant infected with bud rot .
2. The fungus *Phytophthora palmivora* produces intercellur, non septate, hyaline mycelium.
3. Sporangia of *Phytophthora palmivora* releases reniform, biflagellate zoospores upon germination.
4. Thick walled, yellowish brown chlamydospores are produced in *Phytophthora palmivora* causing bud rot of coconut.
5. *Phytophthora palmivora* causing bud rot of coconut survives as chamydospores and oospores in crop residues in the soil.

6. Bud rot diseases spread mainly through zoospores and air-borne sporangia .
7. The pathogen *Ganoderma lucidium* produces fruiting body (Bracket) along the side of the basal trunk in the advanced stages of infection,
8. Primary infection of *Ganoderma lucidium* is through basidiospores in the soil, which attack roots.
9. Drench the soil near the coconut tree with 1 per cent Bordeaux mixture at quarterly interval for thrice a year reduces Ganoderma wilt.
10. Stem injection or root feeding of systemic fungicides Aureofungin- sol or Clixin manages bud rot of coconut.

Answer

S. No	Answer	S. No	Answer
1	True	6	True
2	True	7	True
3	True	8	True
4	True	9	True
5	True	10	True

B. Descriptive questions

a. Long answer questions

1. Describe Ganoderma wilt of coconut under heads, diagnostic symptoms, causal organism, disease cycle and disease management.
2. Which are the important fungal diseases of coconut ? Describe bud rot under heads, diagnostic symptoms, causal organism, etiology, disease cycle, epidemiology and disease management.
3. Describe in detail the causal organism, symptoms, disease-cycle, epidemiology and management of bud rot of coconut.
4. Describe in detail the symptoms, causal organism, etiology, disease-cycle, epidemiology and management of Ganoderma wilt of coconut.
5. Write short notes on

   a. Disese cycle of ganoderma wilt of coconut.

   b. Management of bud rot of coconut.

   c. Etiology of *Phytophthora palmivora* cusing bud rot of coconut.

b. Short answer questions

1. How would you identify the Ganoderma wilt of coconut disease in the field.
2. Write the epidemiological condition for development of bud rot of coconut.
3. Write down the systemic position and etiology of pathogen *Ganoderma lucidum* causing ganoderma wilt of coconut
4. Which fungal pathogen cause bud rot of coconut? Give its systematic position.
5. Elaborate disease cycle of Ganoderma wilt of coconut in pictorial form .
6. Describe in brief about mangment of Ganoderma wilt of coconut.

c. Very short answer questions

1. Discuss epidemiology of Ganoderma wilt.
2. Differentiate between Fusarium wilt and Ganoderma wilt.
3. Name the three fungal diseases of coconut with their causal organism.
4. Name the pathogen or causal organism of coconut disease ; (i) bud rot (ii) ganoderma wilt
5. Write short notes on management practices of Ganoderma wilt.

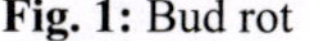

Fig. 1: Bud rot

Fig. 2: Ganoderma wilt

Plate-24: Photograph showing symptoms of major diseases of coconut

25

Tea Crop Diseases & Management

Name of Disease	Causal organism
Blister blight	*Exobasidium vexans*
Red rust	*Cephaleurus mycoidea*
Black rot	*Corticium invisum* and *C. theae*
Black root	*Rosellina areuata*

1) Blister Blight

Diagnostic Symptoms

- Initially oily, yellowish, translucent spots appear on the tender leaf and turn to deep red shiny blisters.
- The circular spot gradually enlarges to 3 to 13mm diameter, bulged on the under surface of the leaf with a concave trough like depression on the upper surface.
- Leaves become curled and distorted.
- First flush of 2-3 young leaves are attacked and the young shoots and buds are killed.
- Mature leaf is not affected.
- In nursery infection, seedlings are stunted with many thin stems instead of a single stalk.
- Repeated attacks cause death of seedlings.
- Badly affected nurseries will have to be abandoned.
- Succulent leaves and green shoots of newly pruned tea are most susceptible (Fig.1).
- Basidiospores cause secondary infection.

Etiology

- Pathogen :*Exobasidium vexans*

- The mycelium is confined to the blistered areas on the leaves.
- They are septae and collect in bundles below the lower epidermis.
- Later by rupturing the epidermisa continuous layer of vertical hyphae are projected on the surface of spot.
- The fungus produces two kind of spores viz., the conidia and basidiospores.
- The conidia are most abundant, borne singly at the tip of long stalks.
- Basidia are formed on the surface in large number but never form a continuous hymenium.

Disease Cycle

- The fungus completes its life cycle in 11-28 days and several generations of spores are produced in a season.
- It produces conidia and basidiospores in the same blister.
- Spores are air borne.
- The basidiospores germinate on tea leaves by forming appresorium from which infection takes place through an infection hypha.
- It is a polycyclic diseases and the secondary infections are repeated till plants are available in the field.
- The perpetuation of the fungus appears to be form the pre existing infected bushes.

Epidemiology

- Relative humidity > 83% for 7 to 10 days favours disease.
- Temperature above 35°C inhibit the disease.
- Bushes in low, moist and shady localities suffer more.
- Pruned trees with new flush is highly susceptible.

Management

- Seedlings should be protected in nursery by weekly sprays of COC@0.3%
- Spray, a mixture of 210g COC + 210g Nickel chloride per ha at 5 days intervals from June-September and 11 day intervals in October-November.

- Mancozeb (0.1%), Tridemorph (0.1%), Triadimefon (0.1%) and Pyracarbolid (Sicarol) offers good disease control under field conditions.

Model Question Paper

A. Objective Type Questions

a. Choose the correct answer from the following

(1) *Exobasidium vexans*belongs to order.

a. Exobasidiales b. Peronosporales
c. Melanconiales d. Moniliales

(2) Blister blight is caused by

a. *Hemileia vastatrix* b. *Rhizoctonia bataticola*
c. *Exobasidium vexans* d. *Sphacelotheca cruenta*

(3) Blister blight of tea of tea is a type of disease

a. Monocyclic b. Polycyclic
c. Polyetic d. None

(4) The fungus *Exobasidium vexans* completes its life cycle in how many days:

a. 11-28 b. 28-35
c. 10-15 d. None

(5) Blister blight disease is

a. Soil borne b. Seed borne
c. Air borne d. All

Answer

S. No	Answer	S. No	Answer
1	a. Exobasidiales	4	a.11-28
2	c. *Exobasidium vexans*	5	c.Air borne
3	b. Polycyclic		

b. State whether the following statements are True or False

1. Mature leaf of tea is mostly not affected with blister blight.
2. Basidiospores of *Hemiliea vastatrix* cause secondary infection in tea plant.

3. Succulent leaves and green shoots of newly pruned tea are most susceptible to blister blight of tea.
4. The fungus*Hemiliea vastatrix*produces two kind of spores *viz.*, the conidia and basidiospores.
5. Relative humidity greater than 83% for 7 to 10 days favours blister blight disease.
6. Temperature above 35°C inhibit the blister blight disease.
7. The fungus*Hemiliea vastatrix*produces conidia and basidiospores in the same blister of tea plant.
8. *Hemiliea vastatrix*spores are air borne in nature.
9. Blister blight of tea is a polycyclic diseases and the secondary infections are repeated till plants are available in the field.
10. The perpetuation of the fungus *Hemiliea vastatrix*appears to be from the pre existing infected bushes.
11. Pruned trees with new flush is highly susceptible to blister blight disease.

Answer

S. No	Answer	S. No	Answer
1	True	6	True
2	True	7	True
3	True	8	True
4	True	9	True
5	True	10	True
		11	True

B. Descriptive Questions

a. Long answer questions

1. Explain in detail about symptoms, etiology, disease cycle and management of blister blight of tea
2. Describe the blister blight of tea in the following headings:

   (i) Pathogen (ii) Symptoms

   (iii)Disease cycle (iv) Disease management
3. Describe in detail the most distinguishing symptoms, causal organism, disease cycle , epidemiology and management of blister blight of tea.

b. Short answer questions

1. Why the blister blight of tea is historically important in India ?
2. Write down the etiology of blister blight of tea.
3. Which fungal pathogen causes blister blight of tea? Give its systematic position.
4. Write down the management practices of blister blight of tea.
5. Describe the causal organisms and diagnostic symptoms of blister blight of tea.

Fig. 1: Blister blight

Plate-25: Photograph showing symptoms of major diseases of tea

26

Coffee Crop Diseases & Management

Name of Disease	Causal organism
Rust	*Hemileia vastatrix*
Black rot or Koleroga	*Cortieium salminicolor*
Die back or Anthranose	*Collectorichum coffeanum*
Damping off / Collar rot	*Rhizocotonia solani*
Berry blotch	*Cercospora coffeicola*

1) Rust

Diagnostic Symptoms

- The disease is restricted to leaves although sometimes it can seen on tender shoots and berries.
- The rust pustules appear as small spots, 1-2 mm in diameter.
- In early stages these spots are yellowish becoming orange in color with increase in size.
- On the upper surface, opposite the spots on the lower surface , the color is often brownish.
- In severe attack leaves may dry and wither.
- The berries remain small and fail to ripen.
- Extensive defoliation weakens the trees and results in poor yield, severe die back of twigs, and death of trees (Fig.1).

Etiology

- Pathogen: *Hemileia vastatrix.*
- The mycelium is intercellular and sends haustoria into the cells.
- The mycelium sends out erumpent stalks through stomata which bear the uredospores.
- The uredospores are reniform or orange segment like in shape.

- The convex side of the spores are echinulated and the lower side is smooth and measure 26 – 40 x 20 – 30 micron meter.
- The telial stage succeeds the uredial stage in the later stage.
- The pycnial and aecial stages of *H. vastatrix* has not been seen and it is unknown whether the fungus is autoecious or heteroecious.

Disease Cycle

- The fungus survives primarily as mycelium, uredia and uredospores on infected leaves.
- The uredospores are easily disseminated by wind, rain and possibly insects.
- Under favourable conditions the spores germinate and enter the host through stomata.
- The mycelium grow intercellularly sending haustoria into the cells.
- Repeated crops of uredospores are produced during the growing season and, as a result, several secondary infection cycles take place if environmental conditions are favourable.
- Young leaves are more susceptible than older leaves.

Epidemiology

- Rainy weather, heavy dew and high temperature favours disease incidence.
- Uredospores germinate only in free moisture.

Management

- Fallen leaves should be composted or destroyed.
- Spray the bushes once with Bordeaux mixture @0.5% or Copper oxy chloride @0.25% and subsequently 2-3 times during monsoon.
- Spray Triadimefon@0.05%.

Model Question Paper

A.Objective Type Questions

a. Choose the correct answer from the following

(1) *Hemileia vastatrix* belongs to order.

a. Pucciniales
b. Peronosporales
c. Melanconiales
d. Moniliales

(2) Rust of coffee is caused by

a. *Hemileia vastatrix*
b. *Rhizoctonia bataticola*
c. *Sphacelotheca sorghi*
d. *Sphacelotheca cruenta*

(3) The uredospores of *Hemileia vastatrix* are like:

a. Orange segment
b. Reniform
c. Both
d. None

(4) The uredospores are easily disseminated by

a. Wind
b. Rain
c. Insects
d. All

(5) Several secondary infection cycles of rust are produced during the growing season as a result of

a. Uredospore
b. Teliospore
c. Basidiospore
d. Aeciospore

Answer

S. No	Answer	S. No	Answer
1	a Pucciniales	4	d.All
2	a *Hemileia vastatrix*	5	a. Uredospore
3	c Both	6	

b. State whether the following statements are True or False

1. The mycelium of *Hemileia vastatrix* sends out erumpent stalks through stomata which bear the uredospores.
2. The uredospores of *Hemileia vastatrix* are reniform or orange segment like in shape.

3. The telial stage succeeds the uredial stage in the later stage of life cycle of *Hemileia vastatrix*.
4. The pycnial and aecial stages has not been reported in of *H. vastatrix*.
5. The fungus *Hemileia vastatrix* survives primarily as mycelium, uredia and uredospores on infected leaves.
6. The uredospores of rust of coffee are easily disseminated by wind, rain and possibly insects.
7. Repeated crops of uredospores are produced during the growing season of coffee due to infection of rust.
8. Young leaves are more susceptible to rust than older leaves of coffee plants.
9. Uredospores of *Hemileia vastatrix* germinate only in free moisture..

Answer

S. No	Answer	S. No	Answer
1	True	6	True
2	True	7	True
3	True	8	True
4	True	9	True
5	True	10	

B. Descriptive Questions

a. Long answer questions

1. Explain in detail about symptoms, etiology, disease cycle and management of rust of coffee.
2. Describe the rust of coffee in the following headings:

   i. Pathogen (ii) Symptoms

   iii Disease cycle (iv) Disease management
3. Describe in detail the most distinguishing symptoms, causal organism, disease cycle , epidemiology and management of rust of coffee.

b. Short answer questions

1. Write down the etiology of rust of coffee.
2. Which fungal pathogen causes rust of coffee? Give its systematic position.

3. Write down the management practices of rust of coffee.
4. Describe the causal organisms and diagnostic symptoms of rust of coffee.
5. Why the rust of coffee is historically important ?

Fig. 1: Rust

Plate-26: Photograph showing symptoms of major diseases of coffee

27

Mango Crop Diseases & Management

Sl.No	Name of Disease	Causal organism
1	Anthracnose	*Colletotrichum gloeosporioides* Ston, Spaull and Schrenk
2	Malformation	*Fusarium mangiferae* Wollenweber & Reinking
3	Bacterial blight	*Xanthomonas campestris* pv. *Mangiferae*
4	Powdery mildew	*Oidium mangiferae* Berthet
5	Stem end rot	*Botrydiplodia theobromae*
6	Red rust	*Cephaleurus mycoides*

Anthracnose

Colletotrichum gloeosporioides (Glomerella cingulata)

Diagnostic Symptoms

- Disease produces symptoms like leaf spot, blossom blight, withertip, twig blight and fruit rot .
- Symptoms on leaves show as gray to brown spots with dark margins and a yellow halo. The spots later enlarge and coaleasce to form sizable necrotic areas.
- Tender shoots and foliage are easily affected which ultimately cause die back young branches.
- Blossom blight may vary in severity from slight to a heavy infection of the panicles .
- Black spots develop on panicles as well as on fruits. Severe infection destroys the entire inflorescence resulting in no setting of fruits.
- Young infected fruits develop black spots, shrivel and drop off.
- Fruits infected at mature stage carry the pathogen into storage and cause considerable loss during storage, transit and marketing (Figure -1).

Etiology

Scientific Classification

Kingdom : Fungi

Division : Eumycota

Sub- division :Deuteromycotina

Class : Coelomycetes

Order :Melanconiales

Family :Melanconiaceae

Genus :*Colletotrichum*

Species :*gloeosporioides*

- *Colletotrichum gloeosporioides* represents the anamorphic state of ascomycetous fungus *Glomerella cingulata*(Telemorphic state).
- Mycelium :septate and slightly dark coloured.
- Hyphae: accumulate below the host cuticle and develop fruiting bodies called acervuli.
- Acervuli: consists of numerous, closely packed , single celled, colorless conidiophores that produce conidia on their tips.It may also posses sterile hair like structures called 'setae'.
- Conidia : broadly oval to oblong with obtuse ends, non septate, some times posses oil globules, measure 12-16x4-6μm.
- Setae : common on twigs but not on fruits.

Disease Cycle

- Pathogen survives in plant debris like twigs and leaves which fall on the ground.
- Primary infection caused by condia developed on disesed plant parts on ground , are disseminated by air , brought on to the susceptible part of plant, germinate in favourable conditions.
- Secondary infection is mainly due to transfer of conidia through rain splash or wind driven rain water.

Epidemiology

- Optimum temperature -25°C.
- Relative Humidity -95-97% for 12 hours.

Management

- Plant resistant varieties and use healthy seedlingsLeave sufficient space between plants.
- Prune trees yearly to enhance ventilation.
- Implement good drainage methods.
- Diseased leaves, twigs, gall midge infected leaves and fruits, should be collected and burnt.
- Cover the fruits on tree, 15 days prior to harvest with news paper or brown paper bags.
- Foliar sprayof Carbendazim(0.1%) or Azoxystrobin 23% SC(0.1%) or Copper oxy-chloride 50% WG (0.25%) at 15 days intervals until harvest.
- Foliar spray of *Pseudomonas fluorescens* @ 5g/liter of water at 21 days interval commencing from october on flower branches.5-7 sprays one to be given on flowers and bunches.
- Treat the fruit with hot water, (50-55°C) for 15 minutes or dip in Thiobendazole solution (1000ppm) for 5 minutes before storage.
- Store fruits in a well ventilated environment.

2) Powdery Mildew

Oidium mangiferae (Acrosporum mangiferae)

Diagnostic Symptoms

- Pathogen parasitizes young tissues of all parts of the inflorescence, leaves and fruits.
- Initially white superficial powdery fungal growth appearon leaves, stalks of panicles, flowers and young fruits.
- Young leaves are attacked on both the sides but it is more evident on the upper surface. Often these patches coalesce and occupy larger areas turning into purplish brown in colour.

- Affected flowers and fruits drop pre-maturely reducing the crop load considerably or might even prevent the fruit set (Figure -4).

Etiology

Scientific Classification

Kingdom : Fungi

Division : Eumycota

Class : Leotiomycetes

Order : Erysiphales

Family : Erysiphaceae

Genus : *Oidium*

Species : *mangifeare*

- Pathogen is an ectoparasitic biotroph and oogamous type.
- Hyphae are superficial, septate, hyaline, and ramify over the host surface forming white dense coating..
- Conidiophores short, hyaline and conidia single celled –barrel shaped, produced in chain.

Disease Cycle

- Pathogen overwinters in dormant bud or on old leaves.
- Resting spores are source of primary infection.
- Secondary spread of disease can occur thorugh air borne conidia produced in these infections.
- Spores blown wind from infected areas readily adhere to hairy, unopened flowers near tip of the inflorescence and germinate in five to seven hours and causes infection.

Epidemiology

- Optimum temperature- 15-30^{0}C.
- Relative humidity-60-90 %.
- Rains or mists accompanied by cooler nights during flowering.
- Most susceptible stage: Full bloom and fruit set at pea size stage.

Management

- Use resistant/tolerant varieties likeNeelum, Zardalu, Banglora, Torapari-khurd and Janardhan pasand.
- Foliar sprays with wettable sulphur@ 0.3% orKarathane@ 0.1% or Azoxystrobin 23% SC @0.1% or Carbendazim 46.27% SC@ 0.1%, Hexaconazole 5% EC@ 0.1% or Triadimefon 25% WP@ 0.040- 0.100% depending on the size of tree.
- Spraying at full bloom needs to be avoided.
- Grow mango on dry and well ventilated areas.
- Intercrop with other non host tree species.
- Prune diseased leaves and malformed panicles harbouring the pathogen to reduce primary inoculums load.

3) Mango Malformation

Fusarium moliliforme var. subglutinans

Diagnostic Symptoms

- Two types of symptoms are generally produced; floral malformation and vegetative malformation.

- **Vegetative malformation**
  - Pronounced in young seedlings.
  - Affected seedlings develop vegetative growths which are abnormal growth, swollen and have very short internodes, giving it a bunch like an appearance on the shoot apex.
  - Seedling remain stunted and eventually die.
- **Floral malformation**
  - Flower buds are transformed into vegetative buds and a large number of small leaves and stems, which are characterized by appreciably reduced internodes and give an appearance of witches" broom.
  - Flower buds seldom open and remain dull green(Figure -2).

Etiology

- Physiological disorder, eriophyid mites, fungal infection, virus, herbicides and other toxic compounds aids in the production of pathogen.
- Deficiency of iron, zinc and copper can also cause the malformation.

Disease Cycle

- Disease is mainly spread via infected plant material.
- Wound by mango bud mite lead to facilitate primary infection.
- Secondary infectionis either by air borne conidia or by conidia carried by eriophid mite *Aceria mangiferae.*

Epidemiology

- Optimm temperature- 10-15^{0}C.
- More disease in young than in old palnts.

Integrated Management

- Use of disease free planting material.
- Monitor orchard regularly for any signs of deformed plants parts.
- Plough the field before planting to destroy existing weeds in the field.
- Deep summer ploughing.
- Irrigate the orchards as and when required.
- Provide proper shade, irrigation & drainage.
- Floral malformed panicles/ vegetative malformed shoots should be pruned and burnt.
- Application of NAA (200 ppm) in the first week of October (Before bud differentiation stages) followed by deblossoming in the late December or January.
- Spray Zinc, Boron and Copper before bloom and after fruit harvesting..

4) Bacterial Blight

Xanthomanas mangiferae- indicae

Diagnostic Symptoms

- Also known as bacterial spot, leaf spot, black spot and mango blight.
- Symptoms appear on all the above plant parts.
- On the leaves, initially slightly water soaked yellowish, translucent irregular spots are formed towards leaf tip which enlarge soon and become dark brown with a yellow halo.These dark spots are limited by veins and become angular resulting in cankerous raised lesions.Spots appear all over the lamina.Under high humid conditions these spots fuse, leaves turn yellow and drop downprematurely.
- On branches, green twigs, and stem, the fresh dark lesions are water soaked, which further become raised and dark brown having longitudinal fissures of infection.The vascular tissues beneath these lesions are filled with gum which oozes out giving a sticky appearance.
- When fruits are attacked, water soaked lesion develop and turn dark brown to black and gradually turn into cankers.Sometimes crack appear on the skin of infected fruits releasing gummy ooze containing bacterial cells.Theseverly affected fruits are shed prematurely(Figure -3) .

Etiology

Scientific Classification

Kingdom : Prokaryotae

Division : Gracilicutes

Class : Proteobacteria

Family : Psudomonadaceae

Genus : *Xanthomonas*

Species : *mangiferae- indicae*

- Gram-ve, rod shaped, non-spore forming.
- Motile with single polar flagellum.
- Colonies are smooth, butyrous, circular, glistening, convex with entiremargin, pale to light yellow in color.
- Yellowish color of the colonies is due to production of characteristic waterinsoluble, non-diffusible pigment.

Disease Cycle

- Bacterium survives on diseased plant parts where it multiplies at early stages.
- Primary infection is through infected leaves, branches, and fruits..
- Secondary infection occurs by wind driven rain and rain splash.
- Pathogen enters mango leaves through stomata and fruits through natural opening and wounds.

Epidemiology

- High temperature-32^{0}C.
- High humidity.

Management

- Use healthy planting material.
- Orhard sanitation and seedling certification.
- Grow Resistant variety viz., Bombay green,Jahagir, Fazari and Suvarnrekha and moderately resistant varieties viz., Langra, Dashehari,Chausa, Bombay,Zardalu,Gulabkhas, Kesar.
- Foliar spray of 0.02% Streptomycin followed by Copper oxychloride (0.3%).
- Ensure good ventilation of the trees.
- Avoid mechanical injury to the mango trees during field work.
- Protect them from strong winds and heavy rains with wind breaks.
- Disinfect working tools and equipments.

Model Question Paper

A. Objective Type Questions

a. Choose the correct answer from the following

(1) *Colletotrichum gloeosporioides* belongs to order.

a. Agonomycetales　　b. Peronosporales

c. Melanconiales　　d. Moniliales

(2) Anthracnose of mango is caused by

a. *Colletotrichum gloeosporioides* b. *Rhizoctonia bataticola*

c. *Sphacelotheca sorghi* d. *Sphacelotheca cruenta*

(3) Powdery mildew of mango is caused by :

a. *Cercospora arachidicola* b. *Oidium mangiferae*

c. *Sphacelotheca sorghi* d. *Sphacelotheca cruenta*

(4) Acervuli of *Colletotrichum gloeosporioides* may posses sterile hair like structures called

a. Oospores. b. Cleistothecium

c. Setae d. Apothecium

(5) *Colletotrichum gloeosporioides* represents the anamorphic state of ascomycetous fungus

a. *Glomerella cingulata* b. *Oidium mangiferae*

c. *Sphacelotheca sorghi* d. *Sphacelotheca cruenta*

(6) Hyphae of *Colletotrichum gloeosporioides* accumulate below the host cuticle and develop fruiting bodies called

a. Acervuli b. Cleistothecium

c. Setae d. Apothecium

(7) Optimum epidemiological condition for development of powdery mildew of mango

a. Optimum temperature- 15-30^{0}C, Relative humidity-60-90 %.

b. Optimum temperature- 15-40^{0}C, Relative humidity-50-70 %.

c. Optimum temperature- 15-50^{0}C, Relative humidity-40-60 %.

d. Optimum temperature- 15-20^{0}C, Relative humidity-40-50 %.

(8) White superficial powdery fungal growth appear on leaves, stalks of panicles, flowers and young fruits of mango is characteristic symptoms of

a. Powdery mildew b. Bud rot

c. Grey blight d. None

(9) *Oidium mangiferae* is

a. Biotroph b. Hemibiotroph

c. Necrotroph d. None

(10) *Oidium mangiferae* belongs to order.

a. Erysiphales b. Peronosporales

c. Melanconiales d. Moniliales

Answer

Sl.No	Answer	Sl.No	Answer
1	c. Melanconiales	6	a. Acervuli
2	a. *Colletotrichum oeosporioides*	7	a. Optimum temperature- 15-300C, Relative humidity-60-90 %.
3	b. *Oidium mangiferae*	8	a. Powdery mildew
4	c. Setae	9	a. Biotroph
5	a. *Glomerella cingulata*	10	a. Erysiphales

b. State whether the following statements are *True* or *False*

1) Secondary infection of anthracnose is mainly due to transfer of conidia through rain splash or wind driven rain water.

2) Hyphae of *Colletotrichum gloeosporioides* accumulate below the host cuticle and develop fruiting bodies called acervuli.

3) Fruits infected with Anthrcnose at mature stage carry the pathogen into storage and cause considerable loss during storage, transit and marketing.

4) Secondary spread of powdery mildew disease can occur thorugh air borne conidia produced in these infections.

5) Secondary infection of mango malformation occur either by air borne conidia or by conidia carried by eriophid mite *Aceria mangiferae.*

6) *Xanthomonas mangiferae-* indicae belong to family Psudomonadaceae.

7) Yellowish color of the colonies of *Xanthomonas mangiferae- indicae* is due to production of characteristic water in soluble, non-diffusible pigment.

8) Foliar spray of 0.02% Streptomycin followed by Copper oxychloride (0.3%) manages bacterial blight of mango.

9) Foliar sprays with wettable sulphur@ 0.3% orKarathane@ 0.1% control powdery mildew of mango.

10) Deficiency of iron, Zinc and Copper can also cause the malformation.

Answer

S. No	Answer	S. No	Answer
1	True	6	True
2	True	7	True
3	True	8	True
4	True	9	True
5	True	10	True

B. Descriptive Questions

a. Long answer questions

a. Describe in detail about symptoms, etiology, disease cycle and management of powdery mildew of mango.

b. Describe in detail the most distinguishing symptoms, causal organism, disease cycle and management of anthracnose of mango.

c. Describe in detail the most distinguishing symptoms, causal organism, epidemiology, disease cycle and management of bacterial blight of mango.

d. Briefly describe the diagnostic symptom and management of powdery mildew and anthracnose disease of mango.

5. Explain the mango malformation in the following headings:

(i) Pathogen (ii) Symptoms

(ii) Disease cycle (iv) Disease management

b. Short answer questions

1) Which bacterial pathogen cause Bacterial blight of mango? Give its systematic position.
2) Elaborate disease cycle of Powdery mildew of mango in pictorial form .
3) Describe in brief about managment of mango malformation.
4) How would you identify the powdery mildew of mango.
5) Write the epidemiological condition for development of powdery mildew of mango.
6) Write down the systemic position and etiology of pathogen *Colletotrichum gloeosporioides causing* Anthracnose of mango.

c. Very short answer questions

1) Discuss about epidemiology of powdery mildew of mango.
2) Differentiate between vegetative and flowral malformtion.
3) Name the three fungal diseases of mango with their causal organism.
4) Name the pathogen or causal organism of following disease ;
   (i) Bacterial blight (ii) Powdery mildew
5) Write short notes on management practices of Powdery mildew.

Fig. 1: Anthracnose

Fig. 2: Malformation

Fig. 3: Bacterial blight

Fig. 4: Powdery mildew

Plate-27: Photograph showing symptoms of major diseases of mango

28

Potato Crop Diseases & Management

Name of Disease	Causal organism
Early blight	*Alternaria solani*
Late blight	*Phytophthora infestans*
Black scurf	*Rhizoctonia solani*
Common scab	*Streptomyces scabies*
Brown rot	*Ralstonia solanacearum*
Wart	*Synchytrium endobioticum*
Black leg (Soft rot)	*Erwinia caratovora subsp. caratovora*
Viral diseases	
Mild mosaic/Interveinal mosaic	Potato virus X
Severe mosaic	Potato virus Y
Leaf roll	Potato leaf roll virus
Potato spindle tuber	Viroid

1. Early Blight

Diagnostic Symptoms

- Brown-black necrotic spot-angular, oval shape characterized by concentric rings .
- The concentric ring produce a target board effect, the most characteristic symptom of the disease.
- Several spot coalesce & spread all over the leaf.
- In severe attacks leaves shrivel and fall down.
- Infection and rotting of tubers has also been reported (Fig.1).

Etiology

Scientific Classification

Kingdom : Fungi

Phylum : Ascomycota

Class : Dothideomycetes

Subclass : Pleosporomycetidae

Order : Pleosporales

Family : Pleosporaceae

Genus : *Alternaria*

Species : *solani*

- Pathogen; *Alternaria solani.*
- Hyphae are light brown or olivaceous which become dark coloured with age.
- The hyphae are branched, septate and inter and intra cellular.
- The coniophores emerge through the stomata or between the epidermal cells.
- The conidia are club shaped with a long beak which is often half the long of the whole conidium.
- The lower part of the conidium is brown while the neck is colorless.
- The body of the conidium is divided by 5 – 10 transverse septa and there may or may not be a few longitudinal septa.

Disease Cycle

- The conidia and the mycelium in the soil or in the debris of the affected plants can remain viable for more than 17 months.
- These conidia or the new conidia found on the overwintered mycelium bring about the primary infection of the succeeding potato crop.
- The conidia formed on the spots developed due to primary infection are disseminated by wind to long distances.
- The conidia from the affected plant may also be disseminated to the adjoining plants by rain and insects.
- Infection occurs through the stomata but direct penetration may also takes place.

Epidemiology

- Dry warm weather with intermittent rain .
- Poor vigor.

- Temperature: 25-30°C.
- Poorly manured crop.

Management

- Summer deep ploughing.
- Soil solarization during summer.
- Field sanitation, rogueing.
- Avoid water logged conditions in the field.
- Follow crop rotation.
- Apply manures and fertilizers as per soil test recommendations.
- Use resistant/tolerant varieties.
- Use healthy, certified and weed seed free tubers.
- Removal and destruction of infected plant debris should be done because the spores lying in the soil are the primary source of infection.
- Spray chlorothalonil 75% WP @ 350-500 g 240-320 l of water/acre (second spray after 14 days interval) or copper oxychloride 50% WP @ 1 Kg in 300-400 l of water/acre or mancozeb 35% SC @ 0.5% or 500 g/100 l water 500 l water or as required depending upon crop stage and equipment used or mancozeb 75% WP@ 600-800 g in 300 l of water/acre or hexaconazole 2% SC @ 1.2 l in 200 l of water/acre (second spray after 21 days interval) or kitazin 48% EC @ 0.20% or 200 ml in 200 l of water or propineb 70% WP @ 300 g in 100 l of water or 0.30% as required depending upon crop stage and plant protection equipment used (second spray after 15 days interval) or zineb 75% WP @ 600- 800 g in 300-400 l of water/acre or captan 70% + hexaconazole 5% WP @ 200- 400 g in 200 l of water/acre (second spray after 21 days interval).

2) Late Blight

Diagnostic Symptoms

- Initially starts from leaf tips or margins and spread inward.
- Small faded green patches on upper surface of leaf which turn into brown spots.
- Downy growth of the pathogen on subsequent lower surface.

- Progressive defoliation and collapse of plants under favourable conditions.
- Water soaked stripes on stem which becomes necrotic.
- Purplish brown spots appear on skin of tubers.
- On cutting, the affected tubers show rusty brown necrosis spreading from surface to the centre (Fig.2).

Etiology

Scientific Classification

Phylum : Heterokontophyta

Class : Oomycota

Order : Peronosporales

Family : Peronosporaceae

Genus : *Phytophthora*

Species : *infestans*

- Pathogen – *Phytophthora infestans.*
- The mycelium is endophytic, coenocytic and hyaline which are inter cellular with double club shaped haustoria type.
- Sporangiophores are hyaline, branched, intermediate and thick walled.
- Sporangia are thin walled, hyaline, oval or pear shaped with a definite papilla at the apex.
- The sporangium may act as a conidium and germinate directly to form a germ tube.
- Zoospores are biflagellate possess fine hairs while the other does not.
- The sexual reproduction in *Phytophthora* is not very common.

Disease Cycle

- Primary infection is through use of infected tubers.
- Mycelium spreads into shoots produced from infected tubers and reaches the aerial parts of the plant.
- Sporangiophore emerges through stomata on stem and leaves and produce sporangia, which are spread by rain to wet potato leaves or stems and cause disease.

- Large number of asexual generation in a growing season kills the foliage rapidly.
- The zoospores found in the soil germinate, penetrate through lentils or wounds into the tubers and send intercellular mycelium and haustoria into the cells and cause infection.

Epidemiology

- High relative humidity (>90%) coupled with suitable temperature are the principal factor governing occurrence of late blight epidemics.
- The optimum temperature for germination of the sporangia by zoospores as 12-13^{0}C and by germ tube 24^{0}C.
- Sporulation can occurs at any temperature from 9-26^{0}C.
- The optimum temperature for growth of mycelium is 16-18^{0}C.

Dutch rules

- Night temperature below the dew point for 4 hours or more.
- Night temperature not below 10°C.
- Cloudiness on the next day.
- Rainfall at least 0.1mm on the following day.

Management

- Regulatory measures.
- Select healthy tubers for planting.
- Delayed harvesting.
- High ridging to about 10-15cm height reduces tuber infection.
- Apply potasic fertilizer.
- Grow resistant varieties such as Kufri Jyothi, Kufri Badshah, Kufri Jeevan, Kufri Sherpa, etc.
- Treat tuber with carbendazim 25% + mancozeb 50% WS @ (1.5 + 3.0) to (1.75 + 3.5) for 10 Kg seed (tuber).
- Spray chlorothalonil 75% WP @ 350-500 g in 240-320 l of water/acre (second spray after 14 days interval) or copper oxychloride 50% WP @ 1 Kg in 300-400 l of water/acre or copper sulphate 2.62 % SC @

400 ml in 200 l of water/acre (second spray after 3 days interval) or cyazafamid 34.5% SC @ 80 ml in 200 l water/acre (second spray after 27 days interval) or dimethomorph 50% WP@ 400 g in 300 l water/acre (second spray after 16 days interval) or mancozeb 75% WG @ 400 in 200 l water/acre (second spray after 3-5 days interval) or mancozeb 75% WP@ 600-800 g in 300 l water/acre or hexaconazole 2% SC @ 1.2 l in 200 l water/acre (second spray after 21 days interval) or mandipropamid 23.4% SC @ 0.2 ml/ l in 200- 300 l of water/acre (second spray after 40 days interval) or cymoxanil 8% + mancozeb 64% WP @ 600- 800 g in 200-300 l of water/acre (second spray after 10 days interval) or famoxadone 16.6% + cymoxanil 22.1% SC @ 200 ml in 200-300 l of water/acre (second spray after 27 days interval) or fenamidone 10% + mancozeb 50% WDG @ 500- 600 g in 200 l of water/acre (second spray after 30 days interval) or metalaxyl M 4% + mancozeb 64% WP @ 025% 1 Kg/ acre in 200-400 l water (second spray after 24 days interval) or metalaxyl 8% + mancozeb 64% WP @ 0.25% 1 Kg/ acre in 400 l water (second spray not less than7 weeks) or metiram 55% + pyraclostrobin 5% WG @ 600-700 g in 200 l water/ acre (second spray after 15 days interval) or azoxystrobin 23% SC@200 ml in 200 l of water/acre.

3. Black Scurf

Diagnostic Symptoms

- Black speck, black speck scab, russet scab on tubers.
- At the time of sprouting dark brown colour appear on the eyes.
- Affected xylem tissue causes to wilting of plants.
- Infected tuber contains russeting of the skin.
- Hard dry rot with browning on internal tissue.
- Spongy mass appear on the infected tuber (Fig.3).

Etiology

- Pathogen: *Rhizoctonia solani.*
- Perfect stage-*Thanetophorus cucumeris.*
- The mycelium is hyaline when young and brown at maturity.
- Hyphae are septate and branched with a characteristic constriction at their junction with the main hyphae.

- The branches arise at a right angle to main axis. Sclerotia are black.
- A basidium bears four sterimata each with a basidiospore at the end.
- The basidiospores are hyaline, elliptical to obovate and thin walled.
- They are capable of forming secondary basidiospores.

Disease Cycle

- The fungus is capable of leading a saprophytic life on the organic material and can remain viable in the soil for several years.
- The sclerotia on the seed tubers is the principal source of infection of the subsequent crop raised with these tubers.
- On return of favourable conditions the mycelium present in the soil may develop producing new hypae.
- The new hypha infects fresh tubers, germinating buds and stem.
- Under adverse condtion, the mycelium turns into sclerotia and serves as means for survival.

Epidemiology

- Moderately cool, wet weather.
- Optimum Temp for infection is 18 °C.
- The black scurf develp more in sandy soil than in clay.

Management

- Disease free seed tubers alone should be planted.
- Soak seed tubers in a solution of trisodium phosphate (90 g/l of water) one day before sowing. The tubers should be thoroughly rinsed and dried in shade.
- Two to three sprays of (streptomycin sulphate 9% + tetracylin hydrocloride 1%) SP @ 40 to 50 ppm solution at an interval of 20 days. First spray 30 days after planting.
- Well sporulated tubrs planted shallow.
- The disease severity is reduced if the land is left fallow for 2 years.

4) Viral Diseases

a) Mild mosaic/Interveinal mosaic

Diagnostic Symptoms

- Often referred as latent potato mosaic.
- Light yellow mottling with slight crinklingon potato plants.
- Interveinal necrosis of top foliage.
- Stunting of diseases plants.
- Leaves may appear slightly rugose where strains of PV Y combines (Fig.4).

Spread

- Potato virus X (PV X).
- Spreads mechanically through rubbing of leaves, contact of infected plants, seed cutting knives, farm implements.
- Root clubbing of healthy and diseased plants in field.

Management

- Disease free seed tubers for planting.
- Rouging of diseased plants.

b) Severe mosaic

Diagnostic Symptoms

- Also called potato leaf drop streak.
- Chlorotic streaks on leaves which become necrotic.
- Necrosis of leaf veins and leaf drop streak.
- Interveinal necrosis and stem/petiole necrosis.
- Plant remain stunted in growth (Fig.5).
- Rugosity and twisting of the leaves occurs in combination with PV X and PV A.

Survival and Spread

- Pathogen: Potato virus Y (PVY).
- Infected tubers.
- Spread by aphids, *Myzus persicae* and *Aphis gossypii.*

Management

- Disease free seed tubers for planting.
- Rouging of diseased plants.
- Aphid control.

c) Leaf roll Symptoms

- Upward rolling of leaves, which have a stiff leathery texture.
- Plants stunted and have a stiff upright growth.
- Phloem necrosis of tubers in some varieties (Fig.6).

Spread

- Pathogen: Potato leaf roll virus.
- Infected seed tubers or by aphids.

Management of viral diseases

- Disease free seed tubers for planting.
- Use healthy seed, hot and cold weather cultivation, green manuring, irrigation, fertilizer application.
- Plant early bulking and/or maturing cultivars to help seed production programme in areas having short aphid-free periods so that the seed crop may escape the population pressure of aphid vectors.
- Apply carbofuran 3% CG @ 6.64 Kg/ acre or thiamethoxam 25% WG @ 40 g in 200 l of water/acre or phorate 10% CG @ 4 Kg/ acre or soil drenching of thiamethoxam 25% WG @ 80 g in 200 l water/ acre for aphid control.

4) Potato spindle tuber – Viroid

Symptoms

- Plants appear erect, spindly and dwarfed.
- Leaves small, erect and leaflets dark green.
- Tubers elongated with tapering ends.
- Tuber eyes are numerous and more conspicuous.

Spread

- Infected seed tubers.
- Mechanically spread by knives used to cut seed tubers.
- Also transmitted by pollen and seed and contaminated mouth parts of grasshoppers, flea beetles and bugs.

Management

- Use of PSTVd free potato seed tubers.
- Disinfestation of cutting knive.

Model Question Paper

Objective Type Questions

A. Choose the correct answer from the following

(1) *Alternaria solani* belongs to order.

a. Pleosporales   b. Peronosporales

c. Hyphomycetales   d. Moniliales

(2) Early blight of potato is caused by

a. *Alternaria solani*   b. *Rhizoctonia bataticola*

c. *Sphacelotheca sorghi*   d. *Sphacelotheca cruenta*

(3) Late blight of potato is caused by

a. *Phytophthora infestans*   b. *Sphacelotheca cruenta*

c. *Rhizoctonia bataticola*   d. *Rhizoctonia bataticola*

(4) The most characteristic symptom of the potato disease like brown-black necrotic spot-angular, oval shape characterized by concentric rings produce a target board effect

a. Alternaria blight *b.* Angular leaf spot

c. Anthracnose d. None

(5) The body of the conidia of *Alternaria solani* is divided by

a. Horizontal septa b. Vertical septa

c. Both d. None

(6) *Phytophthora infestans* belongs to order.

a. Pleosporales b. Peronosporales

c. Hyphomycetales d. Moniliales

(7) Dutch rule is followed for which disease of potato

a. Alternaria blight b. Late blight

c. Black scurf d. None

(8) The optimum temperature for germination of the sporangia by zoospores of *Phytophthora infestans* is.

a. 12-13^{0}C b. 22-24^{0}C

c. 25-30 ^{0}C d. 30-35^{0}C

(9) The optimum temperature for germination of the sporangia by germ tube of *Phytophthora infestans* is.

a. 12 ^{0}C b. 24^{0}C

c. 30 ^{0}C d. 35^0

(10) The perfect stage of *Rhizoctonia solani* is

a. *Ascochyta rabiei* b. *Rhizoctonia bataticola*

c. *Thanetophorus cucumeris* d. *Sphacelotheca cruenta*

(11) The resting structure on the seed tubers is the principal source of infection of the subsequent potato crop raised with these tubers.

a. Sclerotia b. Oospore

c. Chlaymydospores d. Zoospore

(12) Mild mosaic/Interveinal mosaic of potato is caused by .

a. Potato virus X
b. Potato virus Y
c. Potato virus Z
d. Potato virus A

(13) Severe mosaic of potato is caused by .

a. Potato virus X
b. Potato virus Y
c. Potato virus Z
d. Potato virus A

(14) Leaf roll of potato is caused by.

a. Potato virus X
b. Potato virus Y
c. Potato leaf roll virus
d. Potato virus A

(15) Potato spindle tuber is caused by .

a. Potato virus X
b. Potato virus Y
c. Potato virus Z
d. Viroid

(16) Potato leaf roll is transmitted by.

a. Aphid
b. White fly
c. Hopper
d. Termite

Answer

S. No	Answer	S. No	Answer
1	a. Pleosporales	9	*a.* 12 ^{0}C
2	a. *Alternaria solani*	10	c.*Thanetophorus cucumeris*
3	a. *Phytophthora infestans*	11	a.Sclerotia
4	a. Alternaria blight	12	a.Potato virus X
5	c. Both	13	b. Potato virus Y
6	b.Peronosporales	14	c. Potato leaf roll virus
7	a. Alternaria blight	15	d. Viroid
8	a. 12-13^{0}C	16	a. Aphid

(b) State whether the following statements are *True* or *False*

1. The conidia of *Alternaria solani* are club shaped with a long beak which is often half the long of the whole conidium.
2. The conidia or the new conidia found on the overwintered mycelium of *Alternaria solani* bring about the primary infection of the succeeding potato crop.

3. The conidia of *Alternaria solani* formed on the spots developed in early blight of potato due to primary infection are disseminated by wind to long distances.
4. The mycelium *Phytophthora infestans* is endophytic, coenocytic and hyaline which are inter cellular with double club shaped haustoria type.
5. Sporangia of *Phytophthora infestans* are thin walled, hyaline, oval or pear shaped with a definite papilla at the apex.
6. Zoospores of *Phytophthora infestans* are biflagellate possess fine hairs .
7. Rugosity and twisting of the potato leaves occurs in combination with PV X and PV A.
8. The mycelium of *Rhizoctonia solani* turns into sclerotia and serves as means for survival in potato crop under adverse condtion.
9. The black scurf disease in potato develops more in sandy soil than in clay.
10. Severe mosaic in potato is spread by*Myzus persicae* and *Aphis gossypii.*

Answer

S. No	Answer	S. No	Answer
1	True	6	True
2	True	7	True
3	True	8	True
4	True	9	True
5	True	10	True

Descriptive Questions

a. Long answer questions

1. Describe in detail the symptoms, etiology, disease cycle and management of Late blight of potato.
2. Give the diagrammatic representation of disease cycle and epidemiology of late blight and black scurf disease of potato.
3. Briefly describe the diagnostic symptoms and management of any two of the following viral diseases of potato

   (i) Mosaic (ii) Leaf curl

4. Illustrate the Black scurf of potato in the following headings :

   (i) Pathogen (ii) Symptoms

   (ii) Disease Cycle (iv) Disease Management

5. Describe in detail the most distinguishing symptoms, causal organism, disease cycle and management of early blight of potato .

b. Short answer questions

1. Compare between symptom of late blight and early blight of potato .
2. Give the disease cycle and epidemiological factor for development of late blight of potato.
3. Write an integrated management scheduele of late blight of potato.
4. Explain the disease cycle and epidemiological factor of Black scurf of potato.
5. Write systemic position and etiology of pathogen causing early blight of potato.

c. Very short answer Questions

1. Suggest some disease resistant varieties of potato against late blight disease.
2. List some important diseases of potato and their causal organisms.
3. Distinguish between mosaic and leaf curl of potato.
4. Which type of disease is can be effectively managed by crop rotation.
5. Discuss the cultural management practices that should be adhered to in order to manage potato virus diseases.

Fig. 1: Early blight

Fig.2. Late blight

Fig. 3: Black scurf

Fig. 4: Mosaic

Fig. 5: Leaf Curl

Fig. 6: Leaf roll

Plate-28: Photograph showing symptoms of major diseases of potato

29

Tomato Crop Diseases & Management

Name of Disease	Causal organism
Damping Off	*Pythium aphanidermatum*
Early Blight	*Alternaria solani*
Late blight	*Phytophthora infestans*
Fusarium wilt	*Fusarium oxysporum* f. sp. *lycopersici*
Bacterial wilt	*Burkholderia solanacearum*
Mosaic	Tomato mosaic virus (TMV)
Leaf Curl	Tomato leaf curl virus (TLCV)

1. Damping off

Diagnostic Symptoms

- Disease occurs in two stages, i.e. the pre-emergence and the post-emergence phase.
- In the pre-emergence phase, the seedlings are killed just before they reach the soil surface.
- The young radical and the plumule are killed and there is complete rotting of the seedlings.
- The post-emergence phase is characterized by the infection of the young, juvenile tissues of the collar at the ground level.
- The infected tissues become soft and water soaked. The seedlings topple over or collapse (Fig.1)

Etiology

- Pathogen; *Pythium aphanidermatum.*
- The fungus has a characteristic mycelium, which coenocytic, hyaline, freely branching and thick hyphae.
- It reproduces both asexually and sexually.
- The asexual reproduction is through the formation of sporangia.

- The sporangium germinates by giving rise to a tubular structure which ends in a vesicle.
- The zoospores are kidney shaped with two flagella.
- The zoospores germinate to produce a germ tube which gives rise to a characteristic coenocytic mycelium.
- The sexual reproduction of the fungus is characterized by the formation of oogonium, antheridium and oospore.
- Oospore is aplerotic, i.e does not fill the oogonial cell completely.
- The oospores may germinate when conditions are favourable, giving rise to germ tube and mycelium.

Disease Cycle

- The fungus is mainly soil borne.
- The oospores persist in soil, resisting adverse conditions.
- When there is sufficient moisture, they germinate and produce the mycelium which infect the host plant.
- The mycelia stage of the fungus is capable of infection the host plant and multiplying very rapidly.
- The zoospores formed in the vesicles of sporangia are commonly released in soil water to swim about and spread from place to place.
- When sexual reproduction starts, resulting in the formation of oospores which help to survive adverse environmental conditions.
- Thus the fungus is capable of living for many years in soil, completing its life cycle both saprophytically and as a facultative parasite.

Epidemiology

- High humidity, high soil moisture, cloudiness and low temperatures below 24°C for few days .
- Crowded seedlings, dampness due to high rainfall, poor drainage and excess of soil solutes.

Management

- Used raised seed bed.
- Provide light, but frequent irrigation.

- Provide better drainage system.
- Drench with Copper oxychloride 0.3% or Bordeaux mixture 1%.
- Seed treatment with fungal culture *Trichoderma viride* (4 g/kg of seed) or Thiram (3 g/kg of seed).
- Spray 0.2% Metalaxyl when there is cloudy weather.

2. Early Blight

Diagnostic Symptoms

- The disease occurring on the foliage at any stage of the growth.
- Early blight is first observed on the plants as small, black lesions mostly on the older foliage.
- Spots enlarge, and by the time they are one-fourth inch in diameter or larger, concentric rings can be seen in the center of the diseased area.
- Tissue surrounding the spots may turn yellow. If high temperature and humidity occur at this time, much of the foliage is killed.
- Lesions on the stems are similar to those on leaves, sometimes girdling the plant if they occur near the soil line.
- Transplants showing infection by the late blight fungus often die when set in the field. The fungus also infects the fruit, generally through the calyx or stem attachment.
- Lesions attain considerable size, usually involving nearly the entire fruit; concentric rings are also present on the fruit (Fig-2).

Etiology

Scientific Classification

Kingdom : Fungi

Phylum : Ascomycota

Class : Dothideomycetes

Subclass : Pleosporomycetidae

Order : Pleosporales

Family : Pleosporaceae

Genus : *Alternaria*

Species : *solani*

- Pathogen ; *Alternaria solani.*
- Mycelium is septate, branched, light brown which become darker with age.
- Conidiophores are dark colored.
- Conidia are beaked, muriform, dark colored and borne singly.

Disease Cycle

- The disease spreads through infected seeds.
- It may survive in seed bed soils and field, with plant refuse.
- Pathogen already present in soil and plant debris causes primary infection on next crop by wind, rain, irrigation and water splash on the lower leaves.
- Secondary spread occurs through conidia developed on primary spots.
- Infected seed also produce diseased seedlings and serves as primary source.
- Infection takes place through stomata and wounds.

Epidemiology

- Optimum temperature-20-30^{0}C.
- Dry warm weather alternating with the intermittent rains.
- Reduction in plant vigour and senescence.

Management

- Use resistant or tolerant cultivars.
- Change the nursery beds location every season,h eradicate weeds and volunteer tomato plants, fertilize properly.·
- Avoid planting overlapping crops in adjacent area.
- Spray azoxystrobin 23% SC @ 200 ml in 200 l of water/acre or captan 50% WP @ 1000 g in 300-400 l of water/acre or captan 75% WP @ 666.8 g in 400 l of water/acre or copper oxy chloride 50% WP @ 1000 g in 300-400 l of water/acre or copper sulphate 2.62% SC @ 400 ml in 200 l of water/acre or iprodione 50% WP @ 600 g in 200 l of water/acre or kitazin 48% EC @ 80 ml in 80 l of water/acre or mancozeb 35% SC @ 200 g in 200 l water/acre or mancozeb 75% WG @ 400 g in 200 l of

water/acre or pyraclostrobin 20% WG @ 150-200 g in 200 l of water/acre or zineb 75% WP @ 600-800 g in 300-400 l of water/acre or ziram 80% WP @ 600-800 g in 300-400 l of water/acre or famoxadone 16.6% + cymoxanil 22.1% SC @ 200 g in 200 l of water/acre or metiram 55% + pyraclostrobin 5% WG @ 600-700 g in 200 l of water/acre.

Late Blight

Diadnostic Symptoms

- At first water soaked spots appear usually at the edge of the lower leaves.
- In moist weather the spot enlarge and form brown blightened areas with indefinite borders.
- A zone of whitish downy fungal growth can be observed on lesions on the undersurface of leaves. Death of leaves occurs rapidly.
- Petioles and stems are affected with lesions and may die in a short time.
- Purplish brown spots appear on skin of tubers.
- Symptoms on fruits are expressed in the form of dark, greasey spots, which can enlage to occupy the entire fruit. Complete rotting of fruits can be observed (Fig.3).

Etiology

- Pathogen: *Phytophthora infestans.*
- Mycelium is endophytic, coenocytic and hyaline which are inter cellular with double club shaped haustoria type.
- Sporangiophores are hyaline, branched intermediate and thick walled.
- Sporangia are thin walled, hyaline, oval or pear shaped with a definite papilla at the apex.
- Sporangium may act as a conidium and germinate directly to form a germ tube.
- Zoospores are biflagellate possess fine hairs while the other does not.
- Sexual reproduction in *Phytophthora* is not very common.

Disease cycle

- Infected seed and the infected soil may serve as a source of primary infection.

- Infected seeds are mainly responsible for persistence of the disease from crop to crop.
- Secondary infection is caused by air borne sporangia.

Epidemiology

- Wetness.
- Relative Humidity - >90%.
- Moderately warm temperaure.-20-25°C.
- Cool night -12-25°C.

Management

- Select healthy tubers for planting.
- Delayed harvesting.
- High ridging to about 10-15cm height reduces tuber infection.
- Apply potasic fertilizer.
- Treat seed with carbendazim 25% + mancozeb 50% WS @ (1.5 + 3.0) to (1.75 + 3.5) for 10 Kg seed (tuber).
- Spray chlorothalonil 75% WP @ 350-500 g in 240-320 l of water/acre (second spray after 14 days interval) or copper oxychloride 50% WP @ 1 Kg in 300-400 l of water/acre or copper sulphate 2.62 % SC @ 400 ml in 200 l of water/acre (second spray after 3 days interval) or cyazafamid 34.5% SC @ 80 ml in 200 l water/acre (second spray after 27 days interval) or dimethomorph 50% WP@ 400 g in 300 l water/acre (second spray after 16 days interval) or mancozeb 75% WG @ 400 in 200 l water/acre (second spray after 3-5 days interval) or mancozeb 75% WP@ 600-800 g in 300 l water/acre or hexaconazole 2% SC @ 1.2 l in 200 l water/acre (second spray after 21 days interval) or mandipropamid 23.4% SC @ 0.2 ml/ l in 200- 300 l of water/acre (second spray after 40 days interval) or cymoxanil 8% + mancozeb 64% WP @ 600- 800 g in 200-300 l of water/acre (second spray after 10 days interval) or famoxadone 16.6% + cymoxanil 22.1% SC @ 200 ml in 200-300 l of water/acre (second spray after 27 days interval) or fenamidone 10% + mancozeb 50% WDG @ 500- 600 g in 200 l of water/acre (second spray after 30 days interval) or metalaxyl M 4% + mancozeb 64% WP @ 025% 1 Kg/ acre in 200-400 l water (second spray after 24 days interval) or metalaxyl 8% + mancozeb 64% WP @ 025% 1 Kg/ acre in 400 l water

(second spray not less than7 weeks) or metiram 55% + pyraclostrobin 5% WG @ 600-700 g in 200 l water/ acre (second spray after 15 days interval) or azoxystrobin 23% SC@200 ml in 200 l of water/acre.

3. Fusarium Wilt

Diagnostic Symptoms

- The first symptom of the disease is clearing of the veinlets and chlorosis of the leaves.
- The younger leaves may die in succession and the entire may wilt and die in a course of few days. Soon the petiole and the leaves droop and wilt.
- In young plants, symptom consists of clearing of veinlet and dropping of petioles. In field, yellowing of the lower leaves first and affected leaflets wilt and die.
- The symptoms continue in subsequent leaves. At later stage, browning of vascular system occurs.
- Plants become stunted and die (Fig.4).

Etiology

- Pathogen ;*Fusarium oxysporum f. sp. lycopersici.*
- Mycelium is septate and hyaline.
- They produce macro and micro conidia.
- Micro conidia are one celled, hyaline, ovoid to ellipsoid.
- Chlamydospores are formed in older mycelium.
- Three races of pathogen, Races 1,2, and 3, have been identified.

Disease Cycle

- The fungus is seed borne and soil borne.
- The fungus survives in the soil as chlamydospores or as saprophytically growing mycelium in infected crop debris for more than 10 years.
- One of the chief methods of its distribution is by seedlings raised in infected soil.
- Transfer of conidia through surface drainage water and agricultural implements also help in distribution of the pathogen from field to field

Epidemiology

- Optimum temperature-28^{0}C.
- Other conditions which predispose the plant to wilt are short day length, low light intensity, low nitrogen and phosphorus and high potassium.

Management

- Summer ploughing.
- The affected plants should be removed and destroyed.
- Seed treatment with carbendazim (2.5g/kg seed).
- Seed treatment with Trichoderma viride 1% WP @ 9 g/kg seed·
- Root zone application: Mix thoroughly 2.5 kg of the *T. viride* 1% WP in 150 kg of compost or farmyard manure and apply this mixture in the field after sowing/ transplanting of crops.
- Spot drench with Carbendazim (0.1%).
- Crop rotation with a non-host crop such as cereals.

5. Bacterial Wilt

Diagnostic Symptoms

- Characteristic symptoms of bacterial wilt are the rapid and complete wilting of normal grown up plants.
- Lower leaves may drop before wilting. Pathogen is mostly confined to vascular region; in advantage cases, it may invade the cortex and pith and cause yellowbrown discolouration of tissues.
- Infected plant parts when cut and immersed in clear water, a white streak of bacterial ooze is seen coming out from cut ends.

Etiology

- Pathogen; *Burkholderia solanacearum.*
- The bacterium is gram negative, rod shaped often occurs in pairs, motile with 1 – 4 flagella..

Disease cycle

- The bacterium is soil borne, survives for 3 years in fallow and for a unlimited period in cultivated land..

- Infection always occurs through wounds caused by transplanting, cultural operation or nematode infections.
- The organism first moves into the large xylem vessels.
- When a single lateral bundle is invaded drooping of leaves is common but if all bundles are invaded the plant wilt.
- The Chilli, egg plant, grount nut, potato and tobacco are alternative hosts which help the pathogen to survive between tomato crops.

Epidemiology

- Relatively high soil moisture and soil temperature favour disease development.
- The optimum temperature for growth is 30 - 37°C.

Management

- Avoid damage to seedling while transplanting.
- Apply bleaching powder @ 10kg/ha.
- Crop rotations, viz., cowpea-maize-cabbage, okra-cowpea-maize, maize- cowpea-maize and finger millet-egg plant are reported effective in reducing bacterial wilt of tomato.
- Rotate with non-host crops, particularly with paddy·
- Use seedlings from pathogen free seed beds.
- Restriction of irrigation water flowing from affected field to healthy field.

6. Mosaic

Diagnostic Symptom

- The disease is characterized by light and day green mottling on the leaves often accompanied by wilting of young leaves in sunny days when plants first become infected.
- The leaflets of affected leaves are usually distorted, puckered and smaller than normal.
- Sometimes the leaflets become indented resulting in "fern leaf" symptoms.

- The affected plant appears stunted, pale green and spindly.
- The virus is spread by contact with clothes, hand of working labour, touching of infected plants with healthy ones, plant debris and implements (Fig.5).

Etiology

- Pathogen *Tomato mosaic virus* (TMV)
- Virus particles are rod shaped, not enveloped, usually straight with helical construction and measure 300x 8nm.
- The single stranded RNA constitutes about 5 percent of the particle weight.
- Thermal inactivation point is 85-90°C.

Mode of Spread and Survival

- The virus is seed borne and upto 94% of seeds may contain the virus.
- The virus infection occurs during transplanting. It is readily transmissible.
- Many solanaceous plants are susceptible to tomato mosaic virus.
- The virus is spread easily by man and implements in cultural operations or by sap or by animals and by leaf contact.
- The virus particles get entry in the cells after contact and induce symptoms.

Management

- Seeds from disease free healthy plants should be selected for sowing.
- Soaking of the seeds in a solution of Trisodium Phosphate (90 g/litre of water) a day before sowing helps to reduce the disease incidence.
- In the nursery, all the infected plants should be removed carefully and destroyed.
- Seedlings infected with the viral disease should not be used for transplanting.
- Crop rotation with crops other than tobacco, potato, chilli, capsicum, brinjal, etc. should be undertaken.

7. Leaf curl

Diagnostic Symptoms

- Leaf curl disease is characterized by severe stunting of the plants with downward rolling and crinkling of the leaves. The newly emerging leaves exhibit slight yellow colouration and later they also show curling symptoms.
- Older leaves become leathery and brittle. The nodes and internodes are significantly reduced in size.
- The infected plants look pale and produce more lateral branches giving a bushy appearance. The infected plants remain stunted (Fig.6).

Etiology

- Pathogen: *Tomato leaf curl virus* (ToLCV)
- The particles of the virus have a geminate structure, that is, they are made up of two incomplete icosahedral(isometric) particles.
- It is a gemini virus.
- The nucleic acid is a circular DNA molecule.
- The virus particles are 80nm in diameter.

Disease cycle

- It is neither seed nor sap transmissible.
- But seeds from fresh fruits having infection may have the virus on the seed coat.
- The virus is transmitted by white fly, *Bemisia tabaci* and grafting.
- Even a single viruliferous insect is able to transmit the virus

Management

- Raising nursery in protected condition (with net of sufficient mesh size to prevent the entry of vector, whitefly).
- Seeds from disease free healthy plants should be selected for sowing.
- In the nursery, all the infected plants should be removed carefully and destroyed. Seedlings infected with the viral disease should not be used for transplanting.

- Keep yellow sticky traps @ 12/ha to monitor the white fly.
- Raise barrier crops-cereals around the field.
- Removal of weed host. Protected nursery in net house or green house.
- Before transplanting dip the roots of seedlings for 15 minutes in imidacloprid 17.8 % SL @ 60-70 ml in 200 l of water/acre for management of leaf curl vector

Model Question Paper

Objective Type Questions

A.Choose the correct answer from the following

(1) The sexual reproduction of the *Pythium aphanidermatum* is characterized by the formation of

a. Oogonium
b. Antheridium
c. Oospore
d. All

(2) Damping off disease of tomato is caused by:

a. *Pythium aphanidermatum*
b. *Sphacelotheca cruenta*
c. *Rhizoctonia bataticola*
d. *Rhizoctonia solani*

(3) Late blight of tomato is caused by

a. *Phytophthora infestans*
b. *Sphacelotheca cruenta*
c. *Rhizoctonia solani*
d. *Rhizoctonia bataticola*

(4) *Fusarium oxysporum* f. sp. *lycopersici* belongs to order.

a. Pleosporales
b. Peronosporales
c. Hyphomycetales
d. Moniliales

(5) Bacterial wilt of tomato is caused by

a. *Burkholderia solanacearum*
b. *Xanthomonas solanacearum*
c. *Erwinia carotovora*
d. *Bacilus subtilis*

(6) Late blight of tomato is caused by

a. *Phytophthora infestans*
b. *Sphacelotheca cruenta*
c. *Rhizoctonia bataticola*
d. *Rhizoctonia solani*

(7) The most characteristic symptom of the tomato disease like brown-black necrotic spot-angular, oval shape characterized by concentric rings produce a target board effect

a. Alternaria blight
b. Angular leaf spot
c. Anthracnose
d. None

(8) The body of the conidia of *Alternaria solani* is divided by

a. Horizontal septa
b. Vertical septa
c. Both
d. None

(9) *Phytophthora infestans* belongs to order.

a. Pleosporales
b. Peronosporales
c. Hyphomycetales
d. Moniliales

(10) *Alternaria solani* belongs to order.

a. Pleosporales
b. Peronosporales
c. Hyphomycetales
d. Moniliales

(11) Early blight of tomato is caused by

a. *Alternaria solani*
b. *Rhizoctonia solani*
c. *Sphacelotheca sorghi*
d. *Sphacelotheca cruenta*

(12) Late blight of tomato is caused by

a. *Phytophthora infestans*
b. *Sphacelotheca cruenta*
c. *Rhizoctonia bataticola*
d. *Rhizoctonia bataticola*

(13) The most characteristic symptom of the potato disease like brown-black necrotic spot-angular, oval shape characterized by concentric rings produce a target board effect

a. Alternaria blight
b. Angular leaf spot
c. Anthracnose
d. None

(14) The body of the conidia of *Alternaria solani* is divided by

a. Horizontal septa
b. Vertical septa
c. Both
d. None

(15) *Phytophthora infestans* belongs to order

a. Pleosporales
b. Peronosporales
c. Hyphomycetales
d. Moniliales

(16) The optimum temperature for germination of the sporangia by zoospores of *Phytophthora infestans* is.

a. 12-13^{0}C
b. 22-24^{0}C.
c. 25-30 ^{0}C
d. 30-35^{0}C

(17) The optimum temperature for germination of the sporangia by germ tube of *Phytophthora infestans* is.

a. 12 ^{0}C
b. 24^{0}C.
c. 30 ^{0}C
d. 35^{0}C

(18) Chlaymydospore formed in which fungus

a. *Fusarium oxysporum f. sp. lycopersici*
b. *Rhizoctonia bataticola*
c. *Thanetophorus cucumeris*
d. *Sphacelotheca cruenta*

(19) Tomato mosaic virus has

a. ss RNA
b. ds RNA
c. ssDNA
d. ds DNA

(20) Tomato leaf curl virus has

a. ss RNA
b. ds RNA
c. ssDNA
d. Circular DNA

(21) Potato spindle tuber is caused by .

a. Potato virus X
b. Potato virus Y
c. Potato virus Z
d. Viroid

(22) Leaf curl virus of tomato is transmitted by.

a. Aphid
b. White fly
c. Hopper
d. Termite

(A) Answer

S. No	Answer	S. No	Answer
1	d.All	12	a.*Phytophthora infestans*
2	a. *Pythium aphanidermatum*	13	*a.*Alternaria blight
3	*a.Phytophthora infestans*	14	c. Both
4	c. Hyphomycetales	15	b.Peronosporales
5	a. *Burkholderia solanacearum*	16	*a.* 12-13^{0}C
6	a. *Phytophthora infestans*	17	*a.* 12 ^{0}C
7	*a.* Alternaria blight	18	a.*Fusarium oxysporum f. sp. lycopersici*
8	c. Both	19	a.ss RNA
9	b. Peronosporales	20	d.Circular DNA
10	a. Pleosporales	21	d. Viroid
11	a. *Alternaria solani*	22	b. White fly

(b) State whether the following statements are *True* or *False*

1. The seedlings of tomato infected with *Pythium aphanidermatum* topple over or collapse.
2. The zoospores of *Pythium aphanidermatum* are kidney shaped with two flagella.
3. Oospore is *Pythium aphanidermatum* is aplerotic.
4. Damping off disease of tomato is soil borne in nature.
5. Conidia of *Alternaria solani* are beaked, muriform, dark colored and borne singly.
6. Alternaria blight disease spread through infected seed of tomato .
7. A zone of whitish downy fungal growth can be observed on lesions on the undersurface of leaves of tomato is characteristic symptoms of late blight disease of tomato.
8. The mycelium *Phytophthora infestans* is endophytic, coenocytic and hyaline which are inter cellular with double club shaped haustoria type.
9. Sporangia of *Phytophthora infestans* are thin walled, hyaline, oval or pear shaped with a definite papilla at the apex.
10. Zoospores of *Phytlphthora infestans* are biflagellate possess fine hairs .
11. *Fusarium oxysporum f. sp. lycopersici* produces macro and micro conidia.

12. Characteristic symptoms of bacterial wilt of tomato are the rapid and complete wilting of normal grown up plants.

13. Apply bleaching powder @ 10kg/ha reduces the incidence of bacterial wilt of tomato.

14. Tomato mosaic virusis mostly seed borne in nature.

15. Leaf curl disease is characterized by severe stunting of the plants with downward rolling and crinkling of the leaves.

Answer

S. No	Answer	S. No	Answer
1	True	9	True
2	True	10	True
3	True	11	True
4	True	12	True
5	True	13	True
6	True	14	True
7	True	15	True
8	True		

Descriptive Questions

a. Long answer questions

1. Describe in detail the symptoms, etiology, disease cycle, epidemiology and management of Damping off tomato.

2. Describe in detail the symptoms, etiology, disease cycle and management of Fusarium wilt of tomato.

3. Describe in detail the symptoms, etiology, disease cycle and management of Late blight of tomato.

4. Give the diagrammatic representation of disease cycle and epidemiology late blight and early blight disease of tomato.

5. Briefly describe the diagnostic symptoms and management of any two of the following viral diseases of tomato.

   (i) Mosaic (ii) Leaf curl

6. Illustrate the Bacterial wilt of tomato in the following headings

   (i) Pathogen (ii) Symptoms

   (ii) Disease cycle (iv) Disease management

7. Describe in detail the most distinguishing symptoms, causal organism, disease cycle and management of early blight of tomato .

b. Short answer questions

1. Compare between symptom of Late blight and Early blight of tomato.
2. Compare between symptom of Fusarium wilt and Bacterial wilt of tomato.
3. Give the disease cycle and epidemiological factor for development of late blight of tomato.
4. Write an integrated management schedule of bacterial wilt of tomato.
5. Explain the disease cycle and epidemiological factor of Late blight of tomato.
6. Write systemic position and etiology of pathogen causing early blight of tomato.

c. Very short answer questions

1. List some important fungal diseases of tomato and their causal organisms.
2. Distinguish between mosaic and leaf curl of tomato on the basis of symptomatology.
3. Which type of disease is can be effectively managed by crop rotation.
4. Discuss the cultural management practices that should be adhered to in order to manage potato virus diseases.
5. Discuss the management practices that should be adhered to in order to manage tomato Fusarial wilt diseases

Fig. 1: Damping off

Fig. 2: Early blight

Fig. 3: Late blight

Fig. 4: Fusarium wilt

Fig. 5: Mosaic

Fig. 6: Leaf curl

Plate- 29: Photograph showing symptoms of major diseases of tomato

30

Brinjal Crop Diseases & Management

Name of Disease	Causal Organism
Damping off	*Phythium aphanidermatum*
Phomopsis fruit rot or blight	*Phomopsis vexans*
Sclerotina blight	*Sclerotinia sclerotiorum*
Collar rot	*Sclerotium rolfsii*
Bacterial wilt	*Pseudomonas solanacearum*
Little leaf	*Mycoplasma*

1) Phomopsis fruit rot or Blight

Diagnostic Symptoms

- Plants are attacked at all stages of growth.
- On leaves, circular to irregular, clearly defined grayish brown spots having light centers appear. The diseased leaves become yellowish in colour and may drop off.
- Several black pycnidiacan be seen on older spots.
- Lesions on stem are dark brown, round to oval and have grayish centers where pycnidia develop.
- At the base of the stem, the fungus causes characteristic constrictions leading to canker development and toppling of plants.
- On fruits, small pale sunken spots appear which on enlargement cover entire fruit surface. These spots become watery leading to soft rot phase of the disease. A large number of dot like pycnidia also develop on such spots.
- Infection of fruit through calyx leads to development of dry rot and fruits appear black and mummified (Fig.1).

Etiology

Scientific Classification

Kingdom : Fungi

Division : Ascomycota

Class : Sordariomycetes

Order : Diaporthales

Family : Valsaceae

Genus : *Phomopsis*

Species : *vexans*

- Pathogen: *Phomopsis vexans*
- Perfect stage: *Diaporthe vexans*
- Mycelium of the pathogen is septate and hyaline becoming dark with age.
- Pycnidia are submerged and later becoming erumpent with a prominent ostiole.
- Pycnidia with or without beak are found in the affected tissue .They are globose or irregular.
- Conidiophores in the pycnidium are hyaline, simple or branched.
- Conidia are hyaline, one celled and sub cylindrical.
- Conidia are produced on simple to branched conidiophores and are of 2 types: Alpha conidia, which are sub cylindrical and beta conidia, which are filiform and curved. Role of beta-conidia in the epidemiology of the disease is not very clear.
- Ascospores are hyaline, narrowly ellipsoid to bluntly fusoid with one septum.
- Perfect stage produces perithecia in which asci with 8 hyaline, bicelled, ellipsoid-fusoid ascospores are produced which are usually contricted at septum.

Disease Cycle

- Pathogen is seed borne and also survives in plant debris as mycelium and pycnidia.

- Pathogen is locally disseminated as water borne pycniospores.
- Spore exude out through the ostiole and can easily be thrown away by rain drops.They can also disseminated by tools and inscets.
- Inoculum present in the form of mycelium and pycnidia results in the diseased seedlings.

Favourable Conditions

- Wet weather and high temperature.
- Optimum temperature - 26^{0}C.
- Relative humidity- 55%.

Management

- Use of disease free seed.
- Deep summer ploughings.
- Practicing crop rotation.
- Adopting good field sanitation.
- Destruction of infected plant material.
- Seeds obtained from healthy plants should be used for planting.
- Hot water treatment of seed at 50 ^{0}C for 30 minutes.
- Secd treatment with Thiram (2 g/kg seed).
- Foliar spray with Mancozeb (0.2%) or Bordeaux mixture (1%) Thiophanate methyl or Carbendazim @ 0.1% , if required repeat after 15 days.

2) Sclerotina Blgiht

Sclerotinia sclerotiorum (Lib.) De Bary.

Diagnostic Symptom

- Infection may occur at any part of the foliage, mainly the stem or branches.
- A dry and discoloured spot develops at the point of infection.
- It gradually girdles the entire stem and also progresses up and down.

- As a result of tissue necrosis, the portion of the plant beyond the point of infection wilts.
- Damage to brinjal is mainly through partial or complete wilting of the plant.
- Occasionally, the fungus may attack the seedlings in nursery (Fig.1).

Etiology

Scientific Classification

Kingdom : Fungi

Division : Ascomycota

Class : Leotiomycetes

Order : Helotiales

Family : Sclerotiniaceae

Genus : *Sclerotinia*

Species : *sclerotiorum*

- Mycelium is hyaline,much branched, septate hyphae which are inter and intra cellular and invades all tissues of affected host portion.
- Pathogen donot produce true conidia or macroconidia.
- Microsclerotia (Spermatia) produced on sclerotia on the discs of over mature apothecia and in culture, as the food supply declines.
- Spermatia germinate very sparsely in water or culture meida and apparently do not serve as source of infection or dissemination of the pathogen.
- They spermatize the ascogenous cells formed beneath the rind of the sclerotium and thus initiate apothecial development.
- Sclerotia formed on host surface are usually loaf shaped or globose while those formed in the pith of the stem are elongated.
- On germination these scleortia give rise to several stipes which develop the funnel shaped cup at the tip.
- Ascospores are discharged in abundance from these cup.
- Ascospores are hyaline, 1 celled and ovate, always 8 in each ascus which has an apical pore through which spore discharge occurs with violence.

- Pathogen always disseminated by means of ascospores in the absence of asexual reproduction .

Disease Cycle

- Sclerotia survive in soil or in plant debris.
- Pathogen always disseminated by means of ascospores in the absence of asexual reproduction which are the most common structures of infection.

Favarable Conditions

- High soil moisture.
- Optimum temperature- 16 to 21°C.

Management

- Immediately after harvest, the affected plants and debris should be collected and burnt.
- Deep summer ploughing should be given in such a way that surface soil is buried deep.
- Rotation of cropping pattern with crops like beet root, onion, maize, paddy and gingelly eliminate the fungal inoculum in the field.
- Foliar spray with Mancozeb 75 % WP @ 0.25% or Carbandazim 50% WP @ 0.1%, if required repeat after 15 days.

Model Question Paper

Objective Type Questions

A.Choose the correct answer from the following

(1) Phomopsis fruit rot disease of brinjal is caused by :

a. *Phomopsis vexans* b. *Sphacelotheca cruenta*

c. *Rhizoctonia solani* d. *Rhizoctonia bataticola*

(2) Sclerotina blight of brinjal is caused by

a. *Sclerotinia sclerotiorum* b. *Sphacelotheca cruenta*

c. *Rhizoctonia bataticola* d. *Rhizoctonia solani*

(3) *Sclerotinia sclerotiorum* belongs to order

a. Pleosporales b. Peronosporales

c. Helotiales d. Moniliales

(4) Which disease of brinjal is characterized by the development of many dot-like pycnidia on sunken spots that arise and cover the entire fruit surface?

a. Alternaria blight *b.* Angular leaf spot

c. Anthracnose d. Phomopsis fruit rot

(5) *Phomopsis vexans* belongs to order.

a. Pleosporales b. Diaporthales

c. Hyphomycetales d. Moniliales

(6) The perfect stage of fungus *Phomopsis vexans*

a. *Diaporthe vexans* b. *Sphacelotheca cruenta*

c. *Rhizoctonia solani* d. *Rhizoctonia bataticola*

(7) The optimum temperature for development of Phomopsis fruit rot in brinjal is

a. 12 ^{0}C b 22 ^{0}C

c. 26 ^{0}C d. 30 ^{0}C

(8) The optimum temperature for germination of the sporangia by germ tube of *Phytophthora infestans* is.

a. 12 ^{0}C b 24 ^{0}C

c. 30 ^{0}C d. 35 ^{0}C

(9) Microsclerotia formed on sclerotia in culture and on the overmature apothecia when the food supply is reduced

a. *Sclerotinia sclerotiorum* b. *Rhizoctonia bataticola*

c. *Thanetophorus cucumeris* d. *Sphacelotheca cruenta*

(10) *Sclerotinia sclerotiorum* survives in soil as

a. Sclerotia b. Chlymydospore

c. Oospore d. Teliospore

(11) Alpha conidia and beta conidia are produced in which disease of brinjal

a. Alternaria blight   *b.* Sclerotinia blight

c. Anthracnose   d. Phomopsis fruit rot

(A) Answer

S. No	Answer	S. No	Answer
1	a. *Phomopsis vexans*	6	a. *Diaporthe vexans*
2	a. *Sclerotinia sclerotiorum*	7	c. 26 ^{0}C
3	c. Helotiales	8	*a.* 12 ^{0}C
4	d. Phomopsis fruit rot	9	a. *Sclerotinia sclerotiorum*
5	b. Diaporthales	10	a. Sclerotia

(b) State whether the following statements are *True* or *False*

1. Several black pycnidia can be seen on older spots present on brinjal leaf and fruit infected with Phomopsis fruit rot.
2. Brinjals afflicted with *Phomopsis vexans* have pycnidia, either with or without a beak.
3. *Diaporthe vexans* produces perithecia in brinjal in which asci with 8 hyaline, bicelled, ellipsoid-fusoid ascospores are produced .
4. The pathogen *Diaporthe vexans* is seed borne and also survives in brinjal debris as mycelium and pycnidia.
5. *Diaporthe vexans* is locally disseminated as water borne pycniospores.
6. *Sclerotinia sclerotiorum* produces microsclerotia on the discs of over mature apothecia and in culture, as the food supply declines.
7. Ascospores of *Sclerotinia sclerotiorum* are hyaline, 1 celled and ovate, always 8 in each ascus which has an apical pore through which spore discharge occurs with violence.
8. Sclerotia of *Sclerotinia sclerotiorum* survive in soil or in brinjal plant debris.
9. *Sclerotinia sclerotiorum* always disseminated by means of ascospores in the absence of asexual reproduction which are the most common structures of infection.
10. Brinjal sclerotinia blight is controlled by foliar spraying with Mancozeb 75% WP @ 0.25 percent.

C. Answer

S. No	Answer	S. No	Answer
1	True	6	True
2	True	7	True
3	True	8	True
4	True	9	True
5	True	10	True

Descriptive Questions

a. Long answer questions

1. Describe in detail the symptoms, etiology, disease cycle, epidemiology and management of Phomopsis fruit rot of brinjal .
2. Describe in detail the symptoms, etiology, disease cycle and management of Sclerotina blight of brinjal.
3. Briefly describe the diagnostic symptoms and management of any two of the following fungal diseases of brinjal

   (i) Sclerotina blight (ii) Phomopsis fruit rot
4. Illustrate the Phomopsis fruit rot of brinjal in the following headings:

   (i) Pathogen (ii) Symptoms

   (ii) Disease cycle (iv) Disease management
5. Describe in detail the most distinguishing symptoms, causal organism, disease cycle and management of Sclerotina blight of brinjal.

b. Short answer questions

1. Give the disease cycle and epidemiological factor for development of Sclerotina blight of brinjal.
2. Write an integrated management schedule of Sclerotina blight of brinjal.
3. Explain the disease cycle and epidemiological factor of Phomopsis fruit rot of brinjal.
4. Write systemic position and etiology of pathogen causing Phomopsis fruit rot of brinjal.
5. Write systemic position and etiology of pathogen Sclerotina blight of brinjal.

6. Discuss the cultural management practices that should be adhered to in order to manage brinjal fungal diseases.
7. Discuss the management practices that should be adhered to in order to manage Sclerotina blight of brinjal

Fig. 1: Phomopsis fruit rot

Fig. 2: Sclerotina blight

Plate 30: Photograph showing symptoms of major diseases of brinjal

31

Chilli Crop Diseases & Management

Disease	Pathogen
Anthrac nose & Fruit Rot	*Colletotrichum capsici*
Fusarium wilt	*Fusarium oxysporum* f.sp.*capsici*
Powdery mildew	*Leveillula taurica*
Bacterial leaf spot	*Xanthomonas campestris pv. vesicatoria*
Cercospora leaf spot:	*Cercospora capsici*
Viral diseases	
Mosaic	Chilli mosaic virus
Leaf curl	Tobacco leaf curl virus

1. Anthracnose or Fruit Rot and Die Back

Diagnostic Symptoms

The disease occurs in two forms: the die back and the ripe -rot

Die back symptoms

- Small, circular to irregular, brownish black scattered spots appear on leaves.
- Severely infected leaves defoliate.
- Infection of growing tips leads to necrosis of branches from tip backwards.
- Necrotic tissues appear grayish white with black dot like acervuli in the center.
- Shedding of flowers due to the infection at pedicel and tips of branches.

Fruit symptoms

- Ripe fruits are more liable for attack than the green ones.
- Small, circular, yellowish to pinkish sunken spots appear on fruits.
- Spots increase along fruit length attaining elliptical shape.
- Severe infection result in the shrivelling and drying of fruits.

- Such fruits become white or greyish in colour and lose their pungency.
- On the surface of the lesions minute black dot like fruiting bodies called 'acervuli'.
- Concentric rings also develop and fruits appear straw coloured
- The affected fruits may fall off subsequently (Fig.1).

Etiology

Scientific Classification

Kingdom : Fungi

Division : Eumycota

Sub- division : Deuteromycotina

Class : Coelomycetes

Order : Melanconiales

Family : Melanconiaceae

Genus : *Colletotrichum*

Species : *capsici*

- Pathogen: *Colletotrichum capsici*
- The mycelium is septate and inter and intra cellular.
- Conidia in mass appear pinkish.
- Conidia are falcate, fusiform with acute apices and narrow truncated base.
- Acervuli formed are rounded, elongated.
- Setae are brown 1-5 septate , rigid, hardly swollen at the base, slightly tapered towards the paler acute apex.

Disease Cycle

- The pathogen survives in the plant debris, which include infected twigs, stem and fruits.
- The secondary spread take place through the wind borne conidia produced in the infected left out plants.
- The conidia germinate within four hours in water and an appresoria is developed for further growth of the fungus between the plant tissues.

- Infected seeds or badly infected fruits also carry the pathogen and serves as source of primary inoculums.

Epidemiology

- Temp, 28^0 C with RH more than 97%.
- Humid weather with rainfall at frequent intervals.

Management

- Collect and destroy all infected plant parts.
- Collect seeds only form fruits without infection.
- Removal and destruction of Solanaceous weed hosts and infected plant debris.
- Resistant varieties: Perennial, Bengal Green, S20-1, H-4, H-6, Lorai, etc.
- Seed treatment with Thiram or Captan or Mancozeb 2.5g/kg is found to be -effective in eliminating the seed-borne inoculum.
- Foliar spray copper oxy chloride 50% WP @ 1000 g in 300-400 l of water/acre or difenoconazole 25% EC @ 50 ml in 200 l of water/acre or hexaconazole 2% SC @ 1200 ml in 200 l of water/acre or propineb 70% WP @ 200 g in 200-300 l of water/acre or tebuconazole 25% WG @ 200-300 g in 200 l of water/acre or zineb 75 % WP @ 600-800 g in 300-400 l of water/acre or azoxystrobin 23% SC @ 200 ml in 200-300 l of water/acre or kitazine 48% EC @ 80 ml in 200-300 l of water/acre or mancozeb 75% WP @ 600-800 g in 300 l of water or captan 70% + hexaconazole 5% WP @ 200-400 g in 200 l of water/acre.

2. Fusarium Wilt

Diagnostic Symptoms

- Characterised by wilting of the plant and upward and inward rolling of the leaves. The leaves turn yellow and die.
- Commonly appear localised areas of the field where a high percentage of the plants wilt and die, although scattered wilted plants may also occur.
- Disease symptoms are characterised by an initial slight yellowing of the foliage and wilting of the upper leaves that progress in a few days into a permanent wilt with the leaves still attached (Fig.1).

Etiology

Scientific Classification

Kingdom : Fungi

Division : Eumycota

Sub- division : Deuteromycotina

Class : Hyphomycetes

Order : Moniliales

Family : Dematiaceae

Genus : *Fusarium*

Species : *solani*

- Pathogen:*Fusarium oxysporum f.sp.capsici.*
- Mycelium is grayish white.
- Microconidia are formed singly, hyaline and cylindrical.
- Macro conidia are cylindrical to falcate.
- Chlamydospores are globose to oval and rough walled. discoloured, particularly in the lower stem and roots.

Disease Cycle

- The fungus is mainly soil borne and survive in soil mainly as chlaymydospores formed by the hyphal and conidial cells.
- Infection is always through injured roots.
- Deep wounds to expose the xylem help in easy infection.
- After entry into the root, the fungus multiply extensively in vascular tissues, interrupting the transport of nutrients to the above ground plant parts, where the mycelium is formed in abundance to partially or completely plug the vessels, interrupting the transport of nutrients to the above ground plant parts, thus stunting the plant growth.
- Continued interruption of flow of plant nutrients leads to plant wilt.

Epidemiology

- Optimum soil temperature-20-30^{0}C.
- Hot and dry summers followed by rains sees to favor the disease.

Management

- Use of wilt resistant varieties.
- Drenching with 1% Bordeaux mixture or Blue copper or Fytolan 0.25% may give protection.
- Seed treatment with 4g *Trichoderma viride* formulation or 2g Carbendazim per kg seed is effective.
- Mix 2kg *T.viride* formulation mixed with 50kg FYM, sprinkle water and cover with a thin polythene sheet. When mycelia growth is visible on the heap after 15 days, apply the mixture in rows of chilli in an area of one acre.

(3) Viral diseases

a. Mosaic Viruses

Diagnostic Symptoms

- Light green and dark green patches on the leaves.
- These may be sunken or raised (puckering).
- The leaves are generally reduced in size and somewhat become filamentous like.
- Diseased plants produce less flowers and fruits (Fig.1).

Causal Agent

- The virus is sap transmissible and can be easily transmitted through contact during cultivation.
- Natural transmission of virus through aphid vectors such as *Myzus persicae*, *Aphis gossypii*.

Management

- Grow resistant/tolerant varieties.
- Use healthy, certified and weed free seeds.
- Avoid planting overlapping crops in adjacent area.
- In the nursery all the infected plants should be removed carefully and destroyed.

- Ecological engineering of chilli with growing intercrops such as cowpea, maize, coriander etc.
- Rotate the chilli crop with a non-host cereal crop, cucurbit, or cruciferous vegetable.
- Seed treatment with imidacloprid 70% WS @ 400-600 g/100 Kg seed.
- Apply fipronil 5% SC @ 320-400 ml in 200 l of water/acre or oxydemeton methyl 25% EC @ 640 ml in 200-400 l of water/acre or carbofuran 3% CG @ 13320 g/acre or carbosulfan 25% EC @ 320-400 ml in 200-400 l of water/acre or phorate 10% CG @ 4000 g/acre or quinalphos 25% GEL @ 400 g in 200-400 l of water/acre or quinalphos 25% EC @ 400 ml in 200-400 l of water/acre or quinalphos 1.5% DP @ 8000 g/acre.
- Alternate spraying of chemical at 15 days interval till the end of aphid population

b. Leaf Curl

Dianostic Symptoms

- Leaves curl towards midrib and become deformed.
- Stunted plant growth due to shortened internodes and leaves greatly reduced in size.
- Flower buds abscise before attaining full size and anthers do not contain pollen grains.
- The virus is generally transmitted by whitefly. So control measures of whitefly in this regard would be helpful (Fig.1).

Causal agent

- The tobacco leaf curl virus causes the leaf curl disease.
- The virus is not sap transmitted and is also not seed borne.
- In nature, it is transmitted by the white fly (*Bemesia tabaci*).

Management of viral diseases

- Selection of healthy and disease - free seed.
- Nursery beds should be covered with nylon net or straw to protect the seedlings from viral infection.

- Raise 2-3 rows of maize or sorghum as border crop to restrict the spread of aphid vectors.
- Fenpropathrin 30% EC @ 100-136 ml in 300-400 l of water/acre or pyriproxyfen 5% EC + Fenpropathrin 15% EC @ 200-300 ml in 200-300 l of water/acre.
- Collect and destroy infected virus plants as soon as they are noticed.

Model Question Paper

Objective Type Questions

a. Choose the correct answer from the following

(1) Anthrac nose disease of chilli is caused by :

a. *Colletotrichum capsici* b. *Sphacelotheca cruenta*

c. *Rhizoctonia solani* d. *Rhizoctonia bataticola*

(2) Fusarium wilt of chilli is caused by

a. *Fusarium oxysporum f.sp.capsici* b. *Sphacelotheca cruenta*

c. *Rhizoctonia bataticola* d. *Rhizoctonia solani*

(3) *Fusarium oxysporum f.sp. capsici* belongs to order.

a. Pleosporales b. Peronosporales

c. Helotiales d. Moniliales

(4) Minute black dot like fruiting bodies called 'acervuli' produced in which disease of chilli

a. Alternaria blight b. Angular leaf spot

c. Anthracnose d. Leaf spot

(5) *Colletotrichum capsici* belongs to order.

a. Pleosporales b. Diaporthales

c. Hyphomycetales d. Melanconiales

(6) The optimum temperature for development of Anthracnose in chilli is

a. 12^0C b 22^0C.

c. 28 ^{0}C d. 30^0 C

(7) Microsclerotia formed on sclerotia in culture and on the overmature apothecia when the food supply is reduced

a. *Sclerotinia sclerotiorum* b. *Rhizoctonia bataticola*

c. *Thanetophorus cucumeris* d. *Sphacelotheca cruenta*

(8) *Fusarium oxysporum f.sp.capsici* survives in soil as

a. Sclerotia b. Chlamydospore

c. Oospore d. Teliospore

(9) The nature of *Fusarium oxysporum f.sp.capsici is*

a. Soil borne b. Seed borne

c. Air borne d. Vector borne

(10) Leaf curl of chilli is transmitted by.

a. Aphid b. White fly

c. Hopper d. Termite

Answer

S. No	Answer	S. No	Answer
1	*a. Colletotrichum capsici*	6	c.28 ^{0}C
2	*a. Fusarium oxysporum f.sp.capsici*	7	a. *Sclerotinia sclerotiorum*
3	d. Moniliales	8	b. Chlamydospore
4	*c.*Anthracnose	9	a.Soil borne
5	d. Melanconiales	10	b. White fly

b State whether the following statements are *True* or *False*

1. Anthracnose disease occurs in two forms: the die back and the ripe –rot.
2. Minute black dot like fruiting bodies called 'acervuli' develop on the chilli fruit infected with *Colletotrichum capsici*.
3. Setae of *Colletotrichum capsici*are brown 1-5 septate , rigid, hardly swollen at the base, slightly tapered towards the paler acute apex.
4. The secondary spread of anthracnose take place through the wind borne conidia produced in the infected left out chilli plants.
5. The *Fusarium oxysporum f.sp.capsici*is mainly soil borne and survive in soil mainly as chlaymydospores .
6. *Fusarium oxysporum* f.sp. *capsici* belongs to family dematiaceae.

7. Tobacco leaf curl virus is generally transmitted by whitefly.
8. Tobacco leaf curl virus causes the leaf curl disease.
9. Tobacco leaf curl virus is not sap transmitted and is also not seed borne.
10. Natural transmission of Chilli mosaic virus through aphid vectors.

Answer

S. No	Answer	S. No	Answer
1	True	6	True
2	True	7	True
3	True	8	True
4	True	9	True
5	True	10	True

Descriptive Questions

a. Long answer questions

1. Explain the symptoms and management of the following fungal diseases that affect chillies

   (i) Anthracnose (ii) Wilt

2. Use the following headings to illustrate the anthracnose of chilli:

   (i) Pathogen; (ii) Symptoms

   (iii)Disease cycle (iv) Management

3. Give a thorough description of the diagnostic symptoms , causes, disease cycle, epidemiology, and management of Phomopsis fruit rot of brinjal .
4. Clearly describe the symptoms, causes, development of the disease, and managment for chili anthracnose.
5. Give a thorough description of the most distinctive symptoms, the etiological agent, the progression of the disease, and how to manage Fusarium wilt of chili.

b. Short answer questions

1. Describe the epidemiological factors and disease cycle that lead to the development of Fusarium wilt of chilli.
2. Develop an integrated management plan for chiliFusarium wilt
4. Describe the systemic position and etiology of the organism that causes chili anthracnose.

5. Describe the systemic position and etiology of the organism that causes chili Fusarial wilt
6. Explain about the cultural management procedures that need to be followed to control fungal disease in chilli.
7. Discuss about the management procedures that need to be followed to control Fusarium wilt in chilli.

Fig. 1: Anthracnose& Fruit rot.

Fig. 2: Fusarium wilt

Plate-31: Photograph showing symptoms of major diseases of chilli

32

Cucurbits Crop Diseases & Management

Name of Diseases	**Causal organism**
Downy Mildew	*Pseudoperonospora cubensis* (Berkeley & M. A. Curtis)
Powdery mildew	*Erysiphe cichoracearum DC, Sphaerotheca fuligenia* (Schltdl.),
Fusarium wilt	*Fusarium oxysporum* Schlecht
Anthracnose	*Colletotrichum orbiculare*
Cercospora leaf spot	*Cercospora citrullina, C. melonis, C. lagenarium*
Alternaria Blight	*Alternaria cucumerina,*

1) Downy Mildew

Diagnostic Symptoms

- Yellow, angular spots restricted by veins resembling mosaic mottling appear on upper surface of leaves .
- Corresponding lower surface of these spots shows a purplish downy growth in moist weather.
- Spots turn necrotic with age.
- Diseased leaves become yellow and fall down .
- Diseased plants get stunted and die .
- Fruits produced may not mature and have a poor taste (Fig.-1).

Etiology

Scientific Classification

Kingdom : Fungi

Division : Eumycota

Class : Oomycetes

Order : Peronosporales

Family :Peronosporaceae

Genus :*Pseudoperonospora*

Species : *cubensis*

- Pathogen: *Pseudoperonospora cubensis*
- Obligate parasite.
- Mycelium: coenocytic and intercellular with small ovate or finger like haustoria.Mycelium develop one to five sporangiosphores arise through the stomata.
- Sproangiophores: 200-300 µm long and 5-9 µm broad, dichotomous branching at acute angle on which sporangia borne.
- Sporangia: grayish to olivaceous purple, ovoid to ellipsoidal, thin walled with a distal papilla at distal end, measure 21-39x 14-23 µm and germinate producing biflagellate zoospores.
- Zoospores: 10 – 13 micron meter, biflagellate,
- Oospores : not common, spherical, rarely avoid to ellipsoid, light yellow and smooth walled and measure 19-22µm in diameter.

Disease Cycle

- Pathogen survives on the diseased plant debris or wild cucurbits and cause infection through roots.
- Primary infection occurs thorugh fungal growth and sporangia produced on infected plant parts.
- Secondary infection occurs thorugh sporangia germinates and produce zoospores.

Epidemiology

- Relative humidity > 90%.
- High soil moisture.
- Frequent rains.

Management

- Deep ploughing of fields during summer.
- Soil solarization: Cover the beds with polythene sheet of 45 gauge (0.45 mm) thickness for three weeks before sowing for soil solarization which will help in reducing the soil borne pests.

- Apply Trichoderma spp. @ 2.5 kg/acre along with FYM.
- Trellising cucumbers.
- Avoiding overhead irrigation or irrigating only in the late morning hours will limit the amount of time that leaves are wet.
- Control alternate weed hosts (wild cucumber, golden creeper and volunteer cucumbers) in neighbouring fence rows and field edges.
- Foliar spray of Zineb 75% WP @ 0.2% or Cymoxanil 8% + Mancozeb 64% WP @ 0.25%, if required repeat after 10-15 days.

2) Powdery Mildew

Host range

- Pumpkins, Bottle gourd, Coccinia, Cucumber, Ridge gourd, Bitter gourd.

Diagnstic Symptoms

- Whitish or dirty grey, powdery growth on foliage, stems and young growing parts.
- Superficial growth ultimately covers the entire leaf area.
- Diseased areas turn brown and dry leading to premature defoliation and death.
- Fruits remain underdeveloped and are deformed (Fig.-2).

Etiology

Scientific Classification

Kingdom : Fungi

Division : Eumycota

Class : Leotiomycetes

Order : Erysiphales

Family : Erysiphaceae

Genus : *Erysiphe/Sphaerotheca*

Species : *chichoracearum/fuliginea*

- Butler in 1918 reported the occurrence of both the organisms on cucurbits.
- Prior to 1958, *E. cichoracearum* was considered to be primary causal organism throughout the world but now *S. fuliginea* is regarded as the principal causal organism. It is also more common and moreaggressive than the former.
- *E. cichoracearum* is present only in the initial stages of the disease.
- The conidia of the two fungi are very similar and difficult to be distinguished from each other, but the presence of fibrosan bodies in the conidia of *S. fulginea* enables a firm identification of the fungus.

Erysiphe chichoracearum

- Mycelium: colorless, profusely branched, spetate mycelia, develops haustoria inside the host cells to obtain nutrients.
- Conidiophores: erect, stout, small and unbranched producing conidia at the apex.
- Conidia: produced in chain, ellipsoidal or barrel shaped,measure 63.8 x 31.9 micron meter, disseminated by air.
- Cleistothecia:dark, spherical with myceloid appendages which contain 10 – 15 asci.
- Ascus: contains two rarely three ascospores which are single celled , hyaline, measuring 20-30x12-18 mm.
- Cleistotheical stage is not very common.

Sphaerotheca fuliginea

The major point where it differs from that of Erysiphe is the sexual stage.

- Produces cleistotheciaeach containing only one ascus, which is broadly elliptic to subglobose measuring 50-80x30-60μm.
- Each ascus contains eight ascospores and each ascospores is ellipsoid to nearly spherical.

Disease Cycle

- Pathogen overwinter or survives on infected plant debris in the soil as cleistotheica.

- Cliestothecia , though formed , are not significant in disease initiation.
- Pathogen has a wide host range and therefore, is always available on one or the other collateral host.
- Conidial stage of the pathogen release conidia for primary infection on the spring or summer sown cucurbits.
- Secondary infection is caused by wind borne conidia produced during primary infection.

Epidemiology

- Optimum temperature- 30^{0}C.
- Dry conditions.

Management

- Plant resistant/ tolerant varieties.
- Increasing air movement inside the canopy.
- Monitor fields regularly to assess the incidence of a disease.
- Foliar spray with Carbendazim 50% WP 0.2%or benomyl 50% WP @ 0.05% or Thiophanate methyl 70% WP @ 0.1% orCalixin 0.1% or Karathane @0.2%.
- Crop rotation with non susceptible crops.
- Apply balanced fertilizers.
- Remove plant residue after harvest.

3) Fusarium wilt

*Fusarium oxysporum*Schlecht

Diagnostic Symptoms

- Initial symptom of the disease is clearing of the veinlets and chlorosis of the leaves.
- Younger leaves may die in succession and the entire may wilt and die in a course of few days.
- Soon the petiole and the leaves droop and wilt.

- In young trailing plant, symptom consists of clearing of veinlet and dropping of petioles. In field, yellowing of the lower leaves first and affected leaflets wilt and die.
- Symptoms continue in subsequent leaves. At later stage, browning of vascular system occurs.
- Plants become stunted and die (Fig.-3).

Etiology

Scientific Classification

Kingdom : Fungi

Division : Eumycota

Sub- division : Deuteromycotina

Class : Hyphomycetes

Order : Hyphomycetales (Moniliales)

Family : Dematiaceae

Genus : *Fusarium*

Species : *solani*

- Pathogen overwinter or survives on infected plant debris in the soil as chlaymydospores.
- Microconidia are formed singly, hyaline and cylindrical.
- Macro conidia are cylindrical to falcate.
- Chlamydospores are globose to oval and rough walled. discoloured, particularly in the lower stem and roots.

Disease Cycle

- Pathogen survives on seeds or plant debris and other host plant parts.
- Primary spread through seed , soil, water, seedling, workers etc.
- Secondary spread through irrigation water, wind from infected plant debris.

Epidemiology

- Relatively high soil moisture and soil temperature.

Management

- Deep ploughing of fields during summer.
- Use pathogen free seeds.
- Soil solarization: Cover the beds with polythene sheet of 45 gauge (0.45 mm) thickness for three weeks before sowing for soil solarization which will help in reducing the soil borne pests.
- Apply Trichoderma spp. @ 2.5 kg/acre along with FYM.
- Remove and destroy the infected plants and plant debris.
- Avoid water stagnation and maintain proper drainage.

Model Question Paper

Objective Type Questions

a. Choose the correct answer from the following

(1) Downy mildew disease of cucurbits caused by

a. *Pseudoperonospora cubensis* b. *Sphacelotheca cruenta*

c. *Rhizoctonia bataticola* d. *Rhizoctonia bataticola*

(2) Fusarium wilt of cucurbits is caused by

a. *Fusarium oxysporum* b. *Sphacelotheca cruenta*

c. *Rhizoctonia bataticola* d. *Rhizoctonia bataticola*

(3) *Fusarium oxysporum belongs to order.*

a. Pleosporales b. Peronosporales

c. Helotiales d. Moniliales

(4) Minute black dot like fruiting bodies called 'acervuli' produced in which disease of chilli

a. Alternaria blight b. Angular leaf spot

c. Anthracnose d. Leaf spot

(5) *Erysiphe cichoracearum belongs to order.*

a. Pleosporales
b. Diaporthales
c. Hyphomycetales
d. Erysiphales

(6) The optimum temperature for development of powdery mildew of cucurbits is

a. 12^0C
b. 22^0C.
c. 24 ^{0}C
d. 30^0 C

(7) Of late, which pathogen is regarded as the principal causal organism of powdery mildew of cucurbits.

a. *Erysiphe cichoracearum*
b. *Sphaerotheca fuligena*
c. *Thanetophorus cucumeris*
d. *Sphacelotheca cruenta*

(8) The presence of fibrosan bodies in the conidia enables a firm identification of the fungus.

a. *Erysiphe cichoracearum*
b. *Sphaerotheca fuligena*
c. *Thanetophorus cucumeris*
d. *Sphacelotheca cruenta*

(9) *Fusarium oxysporum* survives in soil as

a. Sclerotia
b. Chlymydospore
c. Oospore
d. Teliospore

(10) The nature of *Fusarium oxysporum is*

a. Soil borne
b. Seed borne
c. Air borne
d. Vector borne

(11) *Pseudoperonospora cubensis is*

a. Biotroph
b. Necrotroph
c. Hemibiotroh
d All

(A) Answer

S. No	Answer	S. No	Answer
1	a. *Pseudoperonospora*	7	b.*Sphaerotheca fuligena*
2	a. *Fusarium oxysporum*	8	b.*Sphaerotheca fuligena*
3	d. Moniliales	9	b. Chlymydospore
4	c. Anthracnose	10	a.Soil borne

5	d. Erysiphales	11	a.Biotroph
6	d. 30^0 C		

b. State whether the following statements are True or False

1. *Pseudoperonospora cubensis* is an obligate parasite.
2. The occurrence of both *Erysiphe and Sphaerotheca* on cucurbits was reported byButler.
3. The ascus of *Sphaerotheca fuliginea* contains eight ascospores.
4. *Sphaerotheca fuligena* is perfect stage of *Erysiphe cichoracearum.*
5. *Erysiphe chichoracearum*overwinter on infected plant debris in the soil as cleistotheicia.
6. Cucurbit downy mildew can be controlled with foliar spraying Cymoxanil 8% + Mancozeb 64% WP @ 0.25%.
7. *Erysiphe chichoracearum* overwinter or survives on infected plant debris in the soil as cleistotheica.
8. *Fusarium oxysporum* overwinter or survives on infected plant debris in the soil as chlaymydospores.
9. Foliar spray with sulfur fungicide control powdery mildew disease of water melon.
10. The fungicide metalaxyl is used for management of downy mildew disease of cucurbits.

c. Answer

S. No	Answer	S. No	Answer
1	True	6	True
2	True	7	True
3	True	8	True
4	True	9	True
5	True	10	True

Descriptive Questions

a. Long answer questions

1. Give a thorough explanation of the bottle gourd downy mildew's causal organism, symptoms, etiology, disease cycle, and management.

2. Give an outline of the integrated management and diagnostic symptoms of any two of the following cucurbit disease; The powdery mildew; (ii) the downy mildew; (iii) Fusarium wilt.
3. Under the following headings, describe the powdery mildew of cucrbits:
   (i) Pathogen (ii) Symptoms
   (iii) Disease cycle (iv) Disease management.
4. Provide a thorough description of the cucrbit anthracnose disease cycle, causative organism, most identifying symptoms, and integrated management.
5. Provide a diagrammatic representation of the disease cycle for each of the following cucrbit diseases
   (i) Powdery mildew (ii) Downy mildew
   (iii) Fusarium wilt
6. Provide a thorough description of the causative organism, most identifying symptoms, disease cycle and management of Fusarium wilt of cucurbits.

b. Short answer questions

1. Describe the etiology of powdery mildew of bottle gourd.
2. Which fungus is responsible for muskmelons' downy mildew? Describe its systematic position.
3. Discuss an integrated management strategies anthracnose of water melon.
4. Explain the organisms responsible and the cucurbit powdery mildew's penetration mechanism.
5. Make some recommendations for new generation pesticides to control cucurbit downy mildew.

Fig. 1: Downy mildew

Fig. 2: Powdery mildew

Fig. 3: Fusarium wilt

Plate-32: Photograph showing symptoms of major diseases of cucurbits

33

Cruciferous Vegetables Crop Diseases & Management

Major Diseases	**Causal organism**
Alternaria leaf spot	*Alternaria brassicola, A.brassicae*
Black rot	*Xanthomonas campestris pv. campestris*
Cauliflower mosaic	Cauliflower mosaic virus (CaMV)
Club root of crucifers or Finger and toe disease:	*Plasmodiophora brassicae*
Downy mildew	*Peronospora parasitica*
Powdery mildew	*Erysiphe polygoni*
White rust	*Albugo candida (Syn: Cystopus candidus)*
Bacterial soft rot	*Erwinia carotovora*

1. Alternaria leaf spot

Diagnostic Symptoms

- Spots are small, dark coloured.
- They enlarge, soon become circular & 1mm. in diameter.
- Under humid conditions groups of conidiophores will be formed in the spot.
- Spots develop concentric rings.
- Finally the spots coalesce leading to blighting of leaves.
- Pathogen is seed borne and cause shriveling of seeds and poor germination.
- Linear spots also appear on petioles, stems, pods & seeds (Fig.1).

Etiology

Scientific Classification

Kingdom : Fungi

Phylum : Ascomycota

Class : Dothideomycetes

Subclass : Pleosporomycetidae

Order : Pleosporales

Family : Pleosporaceae

Genus : *Alternaria*

Species : *brassicae/ brassicola*

Characters	***Alternaria brassicae***	***Alternaria brassicola***
Colonies	Amphigenous, effused,pale olive, hairy	Effuse, dark-olivaceous brown to dark blackish brown
Mycelium	Immersed	Immersed
Hyphae	Branched, septate, hyaline, smooth, 4-8 microns thick	Branched, septate, hyaline, smooth,1.5-7.5 microns thick
Conidiophores	Arising in groups of 2-10 or more from the hyphae, emerging through stomata	Arising in groups of 2-10 or more from the hyphae, emerging through stomata
Conidia	Solitary occasionally in chains of up to 4, straight or curved, obclavate, rostrate, with 16-19 transverse septa and 0-8 longitudinal septa.75-350 microns long and usually 20-30 microns thick in the broadest part, the beak about one third to half the length of the conidium and 5-9 microns thick.	Mostly in chains of up to 20 or more,straight, nearly cylindrical,usually, tapering slightly towards the apex or obclavate, the basal cell rounded, the beak usually nonexistent, the apical cell being more or less rectangular or resembling a truncate cone ,with 1-11 transverse septa and few but upto 6 longitudinal septa.

Disease Cycle

- **Primary Infection**: Mycelium persisting in the seed or as spores on seed or from debris.
- **Secondary Infection**: Wind or insect borne conidia.
- On infection pathogen becomes subcuticular in leaves. This is followed by colonization of epidermal and mesophyll cells.

Epidemiology

- Optimum temperature- 16-26^{0}C.
- Relative humidity-79-96%.
- Wind velocity-2.5 to 6 km/hr.

Management

- Long rotations (3 years) without crucifer crops or cruciferous weeds such as wild mustard.
- Allow for good air circulation (i.e. wide spacings, rows parallel to prevailing winds, not close to hedgerows).
- Hot water treatment at 50^0c for 30min.
- Seed treatment with Carbendazim+Mancozeb @0.25%.
- Foliar spray with Mancozeb@0.2% or Copper oxy chloride @ 0.3% twice.

2) Black rot

Xanthomonas campestris pv. campestris

Diagnostic Symptoms

- First appear near the leaf margins as chlorotic or yellow (angular) areas
- Yellow area extends to veins & mid rib forming characteristic 'v' shaped chlorotic spots.
- Veins and veinlets turn brown and finally black.
- The vascular blackening extend beyond affected veins to midrib, petiole and stem.
- In advanced stages, infection may reach the roots system and blackening of vascular bundles occur.
- Bacterial ooze can also be seen on affected parts.
- If the infection is early, the plants wilt and die. If the infection is late plant succumb to soft rot &die (Fig.2).

Etiology

Scientific Classification

Kingdom : Prokaryotae

Division : Gracilicutes

Class : Proteobacteria

Family : Psudomonadaceae

Genus : *Xanthomonas*

Species : *campestris pv. campestris*

- Pathogen is gram negative, short rod with rounded ends and non capsulated.
- It occurs singly, rarely in pairs and motile with single polar flagellum.

Disease Cycle

- Bacterium is internally seed and soil borne. It also survives on plant debris
- Geminating infected seeds serve as the primary source of inoculums.
- Bacteria enter the cotyledons through stomata at the marginal sinus and pass from the cotyledons of young leaves and progress systematically throughout the plant.
- Foliage infection through water pores and insect injuries.
- Chief source of local transmission in the field is by irrigation water or wind splashed rain, cultural practices and transplant.

Epidemiology

- Warm and humid weather.
- Optimum temperature - 27-30°C.
- Relative humidity-80-100%.

Integrated Management

- Crop sanitation.
- Crop rotation for 2-3 years with non-cruciferous crops.
- Grow resistant varieties: Cabbage: Cabaret, Defender, Gladiator, Pusa Muktha Cauliflower: Pusa ice, Pusa snow ball-K-I-F, Sel-12.
- Seed treatment with Aureomycin 1000ppm for 30 min.
- Hot water treatment at 50 0c for 30min, for killing seed borne inoculum followed by a 30min dip in streptocycline 100ppm.
- Drenching the nursery soil with formaldehyde 0.5%.
- Spray Agrimycin-100 or Streptocycline-50ppm at transplanting, curd formation and pod formation.

- Crop rotation for 2-3 yrs with non cruciferous crop.
- Application of bleaching powder at 10.0 to 12.5 kg/ha controls the disease.

3. Caulifflower mosaic

Cauliflower mosaic virus (CaMV)

Diagnostic Symptoms

- The only members of the crucifer family that are susceptible to CAMV are cauliflower.
- Chinese cabbage is especially susceptible to contracting CAMV.
- A clearing or chlorosis along leaf veins is one of the systemic symptoms (vein clearing). This is frequently first observed at the leaf base.
- Later, the leaf develops necrotic spots and dark green patches along the veins (vein banding).
- Along with vein cleaning, leaves may form a visually striking mosaic of light and dark green areas.
- Plant become stunted.
- CaMV infection has been linked to internal necrotic spots in cauliflower that has been kept in storage (Fig.3).

Disease Cycle

- Cruciferous weeds or infected brassica crops are the main sources of CaMV inoculum.
- Numerous aphid species, including the green peach aphid, the cabbage aphid, and the false cabbage aphid, are responsible for spreading the virus to the crop.
- After feeding on an infected plant for just one minute, aphids can pick up and spread CaMV.
- Turnip mosaic virus and CaMV co-infection frequently result in symptoms that are more severe than when either virus is present alone.

Epidemiology

- 16- 20°C is the ideal temperature range for plants to display their symptoms.

Control

- Eradicate cruciferous weeds and volunteers, and incorporate crop debris immediately after harvest.
- Isolate transplant beds from commercial crucifer crops.

Model Question Paper

Objective Type Questions

a. Choose the correct answer from the following

(1) Alternaria leaf spot disease of cruciferous vegetables is caused by

a. *Alternaria brassicola*
b. *Sphacelotheca cruenta*
c. *Alternaria solani*
d. *Rhizoctonia bataticola*

(2) Black rot of cruciferous vegetables is caused by

a. *Xanthomonas campestris pv. campestris*
b. *Sphacelotheca cruenta*
c. *Rhizoctonia bataticola*
d. *Pseudomonas campestris pv. campestris*

(3) Caulifflower mosaic is caused by

a. Potato virus X
b. Potato virus Y
c. Potato virus Z
d. Cauliflower mosaic virus

(4) *Alternaria brassicae* belongs to order.

a. Pleosporales
b. Peronosporales
c. Helotiales
d. Moniliales

(5) Spots develop concentric rings in which disease of cauliflower

a. Alternaria blight
b. Angular leaf spot
c. Anthracnose
d. Leaf spot

(6) The body of the conidia of *Alternaria brassicae* is divided by

a. Horizontal septa
b. Vertical septa
c. Both
d. None

(7) The optimum temperature for development of Alternaria leaf spot of crucifers is

a. 8- 12^{0}C b 5-20^{0}C.

c. 16-26 ^{0}C d. 30-35^0 C

(8). 'V' shaped chlorotic spots on leaves is a characteristics symptom of which disease of cabbage

a. Alternaria leaf spot b. Black rot

c. Club root d. Finger and toe

(9) The only members of the crucifer family that are susceptible to CAMV

a. Cauliflower b. Cabbage

c. Kale d. Turnip

(10) Cauliflower mosaic is transmitted by.

a. Aphid b. White fly

c. Hopper d. Termite

(A) Answer

S. No	Answer	Sl.No	Answer
1	a.*Alternaria brassicola*	6	c. Both
2	a.*Xanthomonas campestris pv. campestris*	7	c.16-26 ^{0}C
3	d.Cauliflower mosaic virus	8	b.Black rot
4	a.Pleosporales	9	a.Cauliflower
5	*a.*Alternaria blight	10	a.Aphid

b. State whether the following statements are *True* or *False*.

1. Chinese cabbage is especially susceptible to contracting CAMV.
2. The best temperature range for Cauliflower to show symptoms of mosaic is between 16 and 20°C.
3. *Xanthomonas campestris pv. campestris* is gram negative, short rod with rounded ends and non capsulated with single polar flagellum.
4. Vein bending and vein clearing are the characteristic symptom of cauliflower mosaic disease.
5. Bacterial oozing is another symptoms of *Xanthomonas campestris pv. campestris* infection in affected cauliflower parts.

6. Turnip mosaic virus and Cauliflower mosaic virus co-infection frequently result in symptoms that are more severe than when either virus is present alone.
7. Cauliflower Alternaria blight is controlled with foliar spray containing COC@ 0.3%.
8. Bleaching powder applied at a rate of 10.0 to 12.5 kg/ha minimizes cauliflower black rot disease.
9. The ideal temperature for the disease known as cauliflower black rot to develop is between 27 and 30°C.
10. The distinctive "v"-shaped chlorotic spots that emerge at the leaf edges as chlorotic or yellow (angular) areas that spread to veins and the midribs are indicative of the black rot disease that affects cauliflower.

c. Answer

S. No	Answer	S. No	Answer
1	True	6	True
2	True	7	True
3	True	8	True
4	True	9	True
5	True	10	True

Descriptive Questions

a. Long answer questions

1. Give a comprehensive understanding of the symptoms, causal organism, disease cycle, and management of Alternaria leaf spot cabbage.
2. Outline the integrated management strategy and symptoms for any two of the cruciferous diseases: mosaic, black rot, and alternaria blight.
3. Give a description of the cauliflower black rot disease in each of the following headings:

   (i) Pathogen (ii) Symptoms

   (iii)Disease cycle (iv) Disease management.
4. Give a comprehensive explanation of the causes, symptoms, disease cycle. and integrated management of the Alternaria leaf spot of cabbage.
5. For each of the following cruciferous diseases provide a diagrammatic

representation of the disease cycle; mosaic, black rot, and alternaria blight.

6. Give a comprehensive explanation of the causes, symptoms that can be identified most easily, disease cycle and integrated management of the Cauliflowermosaic.

b. Short answer questions

1. Describe the etiology of Cauliflower mosaic .
2. Which fungus is responsible for Alternaria leaf spot of cabbage? Describe its systematic position.
3. List the integrated management strategies for cauliflower black rot disease.
4. Explain the organisms responsible and the Alternaria leaf spot of cabbage epidemiological condition.
5. Make some recommendations of new generation fungicides to control Alternaria leaf spot of cabbage.

Fig. 1: Alternaria leaf spot

Fig. 2: Black rot

Fig. 3: Cauliflower mosaic

Plate-33: Photograph showing symptoms of major diseases of Cruciferous Vegetables

34

Beans Crop Diseases & Management

S. No	Name of Disease	Causal organism
1	Anthracnose	*Colletotrichum lindemuthianum*
2	Bacterial blight	*Xanthomonas campestris pv phaseoli / Pseudomonas syringae pathovar phaseolicola.*
3	Bean rust	*Uromyces appendiculatus*
4	Yellow mosaic	*Bean yellow mosaic virus, Mungbean yellow virus*
5	Common bean mosaic / Green mosaic	*Bean common mosaic virus (ss RNA)*

1) **Anthracnose**

Diagnostic Symptoms

- All the above ground parts are affected. However, the characteristic symptoms appear on pods.
- On cotyledons, spots are sunken dark brown or black with pink spore mass.
- Seedling infection results in collapse of seedling.
- Spots on leaves appear on lower side and are black. Later these may also appear on upper surface. When the infection is severe, the affected plants wither off.
- Black, sunken, circular spots of varying sizes appear on pods with bright red, yellow or orange margins. The centre of these spots later turns grey or pink due to sporulation of the pathogen. The border of these spots appear raised (Fig.1).

Etiology

- Pathogen: *Glomerella lindemuthianum* (Sacc.& Magn.).
- Conidial stage is *Colletotrichum lindemuthianum.*

- Mycelium is branched, septate, hyaline at first and dark colored with age.
- Acervuli develop beneath the cuticle later rupturing it and becoming erumpent.
- Conidia are borne acrogenously on short conidiophores and appears in pink masses.
- Setae are few, brown and septate,4.2 microns wide and less than 100 microns long.
- Conidia are one celled, hyaline and cylindrical with rounded ends or with one end slightly pointed, measure 13-22x2.5-5.5 microns and germinate by 1-4 germ tubes.
- Appresoria are sparse, pale to dark brown, clavate or circular in outline, regular and 8 x6.7 microns.
- Pathogen produces perithecia, which are 120-210µm in diameter.

Disease Cycle

- Pathogen overwinters in infected seeds and in plant debris.
- Conidia and /or dormant mycelia in the infected seeds germinate / become active and infect the young seedlings.

Epidemiology

- Relative humidity- 92% percent.
- Optimum temperature- 17^{0}C.
- Pathogen requires about 10 mm of rain to establish initial infection.

Management

- Use certified disease-free seed.
- Seed treatment with Carbendazim@2g/kg seed.
- Protect the crop by spraying 0.1% Carbendazim or 0.2%, Zineb or 0.2% Mancozeb at 7-10 days interval.

2. Bacterial Blight

Xanthomonas campestris pv phaseoli.

Diagnostic Symptoms

- Two widespread bacterial blights that affect beans, common blight (*Xanthomonas campestris pv phaseoli)* and halo blight (*Pseudomonas syringae pathovar phaseolicola)*.
- Stems, leaves and fruits of bean plants can be infected by either disease (Fig.2).

Halo blight

- Light greenish-yellow circles that look like halos form around a brown spot or lesion on the plant.
- With age, the lesions may join together as the leaf turns yellow and slowly dies.
- Stem lesions appear as long, reddish spots.
- Rain and damp weather favor disease development.
- Halo blight occurs primarily when temperatures are cool.

Common blight

- Infected leaves turn brown and drop quickly from the plant.
- Infected pods do not have the greenish-yellow halo around the infected spot or lesion.
- Common blight occurs mostly during warm weather.

Etiology

- Gram negative.
- Rod shaped.
- Non capsulated.
- Motile with single polar flagellum.

Disease Cycle

- Pathogen is seed borne.
- Bacteria can live in the soil for two years on plant debris.

- In new area, disease spreads through infected seeds.
- Disease spread through wind splashed rains from diseased to healthy plants.

Epidemiology

Halo blight

- Rain and damp weather.
- Cool temperatures.

Common blight

- Mostly during warm weather.

Integrated Management

- Buy new seeds each year.
- Avoid overhead watering.
- Do not touch plants when the foliage is wet.
- Do not plant beans in the same location more frequently than every third year.
- Apply fixed copper at ten day intervals. Wait one day between spraying and harvest.

Model Question Paper

Objective Type Questions

a. Choose the correct answer from the following

(1) Anthracnose disease of beans is caused by

a. *Colletotrichum lindemuthianum* b. *Sphacelotheca cruenta*

c. *Phytophthora infestans* d. *Rhizoctonia bataticola*

(2) Bacterial wilt of beans is caused by

a. *Xanthomonas campestris pv phaseoli* b. *Sphacelotheca cruenta*

c. *Phytophthora infestans* d. *Rhizoctonia bataticola*

(3) *Alternaria brassicae* belongs to order.

a. Pleosporales
b. Peronosporales
c. Helotiales
d. Moniliales

(4) Acervuli formed in

a. *Erysiphe cichoracearum*
b. *Colletotrichum lindemuthianum*
c. *Pythium aphanidermatum*
d. *Sphacelotheca cruenta*

(5) *Colletotrichum lindemuthianum produces*

a. Sclerotia
b. Chlamydospore
c. Oospore
d. Perithecia

(6) The optimum temperature for development of Anthracnose of beans is

a. 17^0C
b. 25^0C.
c. 30 ^{0}C
d. 35^0 C

(7) Common blight of beans is caused by

a. *Xanthomonas campestris pv phaseoli*
b. *Sphacelotheca cruenta*
c. *Alternaria solani*
d. *Rhizoctonia bataticola*

(8) Halo blight of beans is caused by

a. *Pseudomonas syringae pv phaseolicola*
b. *Sphacelotheca cruenta*
c. *Rhizoctonia bataticola*
d. *Xanthomonas syringae pv phaseolicola*

Answer

S. No	Answer	S. No	Answer
1	a. *Colletotrichum lindemuthianum*	5	d. Perithecia
2	*a. Xanthomonas campestris pv phaseoli*	6	*a.* 17^{0}C
3	a. Pleosporales	7	a.*Xanthomonas campestris pv phaseoli*
4	b. *Colletotrichum lindemuthianum*	8	a.*Pseudomonas syringae pv phaseolicola*

b State whether the following statements are *True* or *False*

1. Conidial stage of anthracnose of beans is *Colletotrichum lindemuthianum.*
2. Acervuli of *Colletotrichum lindemuthianum* develop beneath the cuticle later rupturing it and becoming erumpent.
3. The fungus *Colletotrichum lindemuthianum* produces perithecia.
4. The pathogen *Colletotrichum lindemuthianum* overwinters in infected seeds and in beans debris.
5. Halo blight occurs in beans primarily when temperatures are cool.
6. Common blight occurs in beans mostly during warm weather.
7. *Xanthomonas campestris pv phaseoli* is gram negative, rod shaped, non capsulated, motile with single polar flagellum.
8. Common blight spreadin new area through infected seeds.

Answer

S. No	Answer	S. No	Answer
1	True	5	True
2	True	6	True
3	True	7	True
4	True	8	True

Descriptive Questions

a. Long answer questions

1. Give a comprehensive understanding of the causal organism , symptoms, disease cycle, and management of Anthracnose disease of beans .
2. Give a comprehensive explanation of the causes, symptoms, disease cycle.and management of the bacterial wilt of beans.

3. Depict a diagrammatic representation of the disease cycle of anthracnose, and bacterial wilt disease of beans.

4. Give a comprehensive explanation of the causes, symptoms that can be identified most easily, disease cycle and integrated management of the halo blight of beans.

5. Mention the causes, symptoms that can be identified most easily, disease cycle and integrated management of the bacterial wilt of beans.

b. Short answer questions

1. Describe the etiology of *Colletotrichum lindemuthianum* causing anthracnose of beans.

2. Which fungus is responsible for bacterial blight of beans. Describe its systematic position.

3. List the integrated management strategies for anthracnose of beans..

4. Explain anthracnose of beans epidemiological condition..

5. Make some recommendations for new generation fungicides to control anthracnose of beans.

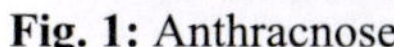

Fig. 1: Anthracnose

Fig. 2: Bacterial blight

Plate-34: Photograph showing symptoms of major diseases of beans

35

Bhendi Crop Diseases & Management

S. No	Name of Disease	Causal organism
1	Vein-Clearing/Yellow vein mosaic	*Bhendi yellow vein mosaic virus*
2	Cercospora leaf spots	*Cercospora malayensis, C. abelmoschi*
3	Fusarium wilt	*Fusarium oxysporum f.sp. vasinfectum*
4	Powdery mildew	*Erysiphe cichoracearum*
5	Little leaf	*Mycoplasma*

1. Vein-Clearing/Yellow Vein Mosaic

Bhendi yellow vein mosaic virus

Diagnostic Symptoms

- Yellowing of the entire network of veins in the leaf blade is the characteristic symptom.
- In severe infections, the younger leaves turn yellow, become reduced in size and the plant is highly stunted.
- The veins of the leaves will be cleared by the virus and intervenal area becomes completely yellow or white.
- In a field, most of the plants may be diseased and the infection may start at any stage of plant growth.
- Infection restricts flowering and fruits, if formed, may be smaller and harder.
- The affected plants produce fruits with yellow or white colour and they are not fit for marketing (Fig.1).
- The virus is spread by whitefly.

Disease cycle

- Primary Infection: Virus particles in infected plants and collateral hosts like *Hibiscus tetraphyllus,Croton sparsiflora* and *Ageratum* spp.
- Secondary Infection: The virus is transmitted by the whitefly, *Bemisia tabaci.*

Management

- Field sanitation, roguing.
- Plant tall border crops like maize, sorghum or pearl millet to reduce whitefly infestations (4 rows).
- Install yellow sticky traps @ 2/acre for monitoring purpose.
- Peppermint plants act as repellent for whitefly.
- French bean acts as an attractant plant for predatory thrips.
- Grow disease tolerant varieties.
- Follow sprinkler type of irrigation.
- Spray NSKE 5% or Azadirachtin 0.03% (300 ppm) neem oil based WSP @ 1000- 2000 ml in 200-400 l of water/acre or azadirachtin 5% W/W neem extract concentrate @ 80 ml in 160 l of water/acre.
- Spray fenpropathrin 30% EC @ 100-136 ml in 300-400 l of water/acre or thiamethoxam 25% WG @ 40 g in 200-400 l of water/acre or pyriproxyfen 5% EC + fenpropathrin 15% EC @ 200-300 ml in 200-300 l of water/acre.

Model Question Paper

Objective Type Questions

a. Choose the correct answer from the following

(1) Yellow vein mosaic disease of bhendi is caused by :

a. Bhendi yellow vein mosaic virus b. Bhendi mosaic virus

c. Tobacco mosaic virus d. Tomato mosaic virus

(2) Yellowing of the entire network of veins in the leaf blade is the characteristic symptom of which disease of bhindi

a. Yellow vein mosaic b. Cercospora leaf spots

c. Powdery mildew d. Fusarium wilt

(3) Yellow vein mosaic is transmitted by.

a. Aphid b. White fly

c. Hopper d. Termite

(4) Powdery mildew of bhindi

a. *Erysiphe cichoracearum*

b. *Colletotrichum lindemuthianum*

c. *Pythium aphanidermatum*

d. *Sphacelotheca cruenta*

Answer

S. No	Answer	S. No	Answer
1	a. Bhendi yellow vein mosaic virus	3	b. White fly
2	a. Yellow vein mosaic	4	a.*Erysiphe cichoracearum*

b. State whether the following statements are *True* or *False*

1. Yellowing of the entire network of veins in the leaf blade is the characteristic symptom of vein-clearing.
2. The bhendi yellow vein mosaic virus is transmitted by the whitefly.
3. The yellow vein mosaic affected plants produce fruits with yellow or white colour and they are not fit for marketing.

Answer

S. No	Answer	S. No	Answer
1	True	3	True
2	True	4	True

Descriptive Questions

a. Long answer questions

1. Give a comprehensive understanding of the causal organism , symptoms, disease cycle, and management of yellow vein mosaic of bhindi.
2. Mention the causes, symptoms that can be identified most easily, disease cycle and integrated management of the yellow vein mosaic of bhindi.

b. Short answer questions

1. List the integrated management strategies for yellow vein mosaic of bhindi..
2. Explain epidemiological condition of yellow vein mosaic of bhindi.

3. Make some recommendations for new generation pesticides to control yellow vein mosaic of bhindi.
4. Explain mode of transmission of Bhendi yellow vein mosaic virus.

Fig. 1: Yellow Vein Mosaic

Plate-35: Photograph showing symptoms of major diseases of bhendi

36

Turmeric Crop Diseases & Management

Name of Diseases	Causal Organism
Leaf spot	*Colletotrichum capsici Syd.*
Leaf blotch	*Taphrina maculans E. J. Butler*
Rhizome rot	*Pythium graminicolum or P. aphanidermatum (Edson)*

1) Leaf Spot

Diagnostic Symptoms

- Initially oblong brown spots having about 4-5 cm in length and 2-3 cm in width with grey centres surrounded with yellow hallo are found on the leaves.
- Large number of spots may be found on a single leaf and as the disease advances, spots enlarge and cover a major portion of leaf blade.
- Severly affected plants dry and wilt.
- Black dots acervuli formed in concentric rings on spot.The grey centres become thin and get teared(Fig-1).

Etiology

Scientific Classification

Kingdom : Fungi

Division : Eumycota

Sub- division : Deuteromycotina

Class : Coelomycetes

Order : Melanconiales

Family : Melanconiaceae

Genus : *Colletotrichum*

Species : *capsici*

- *Pathogen: Colletotrichum capsic*
- Mycelium: septate and inter and intra cellular.
- Conidiophores: single celled ,club shaped arise from the hymenial layer below the epidermis, and emerge directly through the epidermis or through stomata.
- Conidia: The conidiophores bear single conidia which are cylindrical to falcate, single celled, hyaline and mostly have blunt ends. They are densely granular, contain oil globules and measure 18-25x3.5-5µ.
- Acervuli: rounded and elongated.
- Setae: brown 1-5 septate, rigid, hardly swollen at the base, slightly tapered towards the paler acute apex.

Disease cycle

- Pathogen survives in soil and plant debris.
- Primary infection occurs by spores present in soil.
- Secondary infection by conidia through rain or wind.

Epidemiology

- Relative humidity-80%.
- Temperature – 21- 25^0 C.
- Leaf wetness.

Management

- Collection and burning of fallen leaves.
- Proper spacing should be maintained.
- Select seed material from disease free areas.
- Crop rotations should be followed.
- Cultivate resistant/ tolerant varieties: Ashwini, Ambika, Angeles, American pride, Surabhi.
- Treat seed material with Mancozeb at 3 gm/litre of water or Carbendazim at 1 gm/litre, for 30 minutes and shade dry before sowing.
- Foliar sprays with Carbendazim + Mancozeb @ 0.2%,Blue copper or Blitox 50 @ 0.3 percent .

2) Leaf Blotch

Diagnostic Symptoms

- Intial symptom spear as small, oval, rectangular or irregular brown spots on either side of the leaves which soon become dirty yellow or dark brown and leaves also turn yellow.
- In severe cases the plant present a scorched appreance.
- Rhizome yield is reduced(Fig-2).

Etiology

Scientific Classification

Kingdom : Fungi

Division : Ascomycota

Class : Taphrinomycetes

Order : Taphrinales

Family : Taphrinaceae

Genus : *Taphrina*

Species : *maculans*

- Pathogen: *Taphrina maculans.*
- Pathogen hyphae are present in the cuticle and epidermal layers of the host.
- Hausotria: branched or lobedgrow into the host cell, drawing out the required nutrients.
- Asci; are cylindrical to clavate, contains eight ascospores .
- Ascospores: are ovoid, hyaline, unicellular, and measure 6-7x2-3μ.They often multiply by budding either in the asci or when released.

Disease Cycle

- Pathogen survives in soil and plant debris.
- Primary infection occur by spores present in the soil
- Secondary infection by conidia thorugh rain or wind.

Epidemiology

- High soil moisture.
- Temperature 25^0C.
- Leaf wetness.

Management

- Proper spacing should be maintained.
- Select seed material from disease free areas.
- Crop rotations should be followed.
- Cultivate resistant/ tolerant varieties: Ashwini, Ambika, Angeles, American pride, Surabhi.
- Dip seed material with Carbendazim + Mancozeb at 2.5 gm/litre of water for 30 minutes and shade dry before sowing.
- Foliar sprays with Blue copper or Blitox 50 @ 0.3 percent.
- Collection and burning of fallen leaves.

Model Question Paper

a. Choose the correct answer from the following

(1) Leaf spot of turmeric is caused by :

a. *Erysiphe cichoracearum* b. *Colletotrichum capsici*

c. *Pythium aphanidermatum* d. *Sphacelotheca cruenta*

(2) Leaf blotch of turmeric is caused by:

a. *Taphrina maculans*

b. *Colletotrichum lindemuthianum*

c. *Pythium aphanidermatum*

d. *Sphacelotheca cruenta*

(3) Foliar spray with metalaxyl fungicide control which disease of turmeric

a. Rhizome rot b. Leaf spot

c. Powdery mildew d. Leaf blotch

(4) The fruiting body formed in leaf spot of turmeric is called

a. Acervuli b. Pycnidia

c. Apothecia d. Chlaymydospore

(5) The asci of *Taphrina maculans* contains how many ascospores

a. 2 b. 4

c. 6 d. 8

(6) *Taphrina maculans* belongs to order

a. Taphrinales b. Peronosporales

c. Helotiales d. Moniliales

(7) *Colletotrichum capsici* belongs to class

a. Coelomycetes b. Peronosporaceae

c. Pythiaceae d. Moniliaceae

(8) The optimum temperature for development of leaf spot of turmeric is

a. 8- 12^{0}C b. 15-20^{0}C.

c. 21- 25^0 C d. 30-35^0 C

b Answer

S. No	Answer	S. No	Answer
1	b. *Colletotrichum capsici*	5	d. 8
2	a. *Taphrina maculans*	6	a. Taphrinales
3	a. Rhizome rot	7	a. Coelomycetes
4	a. Acervuli	8	c. 21-25^0 C

b. State whether the following statements are True or False

1. *Colletotrichum capsici is a* facultative parasite.
2. Asci of *Taphrina maculans* contains eight ascospores.
3. Setae is one of the identifying character of *Colletotrichum capsici.*
4. An oblong brown spots with grey centres surrounded with yellow hallo is diagnostic symptom of leaf spot of turmeric.
5. Secondary infection takes place by conidia through rain or wind in leaf spot of turmeric.

6. Black dots acervuli formed in concentric rings in spot developed on turmeric leaf due to infection of *Colletotrichum capsici.*
7. Secondary infection of leaf spot disease occur in turmeric by conidia through rain or wind.
8. *Taphrina maculans* belongs to division Ascomycota.
9. Dip seed of rhizome with Carbendazim + Mancozeb at 2.5 gm/litre of water for 30 minutes and shade dry before sowing control leaf spot of turmeric.
10. *Colletotrichum capsici* belongs to the family melanconiaceae.

Descriptive questions

a. Long answer questions

1. Explain in detail the symptoms, etiology, disease cycle and management of rhizome rot of turmeric .
2. Briefly describe the diagnostic symptoms and integrated management oof the following turmeric diseases :

   (i) Leaf blotch (ii) Leaf spot
3. Describe the leaf spot of turmeric in the following headings:

   (i) Pathogen (ii) Symptoms

   (ii) Disease cycle (iv) Management\
4. Describe in detail the most distinguishing symptoms, causal organism, disease cycle, epidemiology and management of leaf blotch of turmeric.
5. Give the diagrammatic representation of the disease cycle of the following diseases of turmeric.

   (i) Leaf blotch (ii) Leaf spot

   (iii)Rhizome rot

b. Short answer questions

1. How will you differentiate leaf spot from leaf blotch of turmeric .
2. Which fungal pathogen causes leaf spot of turmeric? Give its systematic position.
3. Write down an integrated management practices of rhizome rot of turmeirc.

4. Mention etiology of *Colletotrichum capsici* causing leaf blotch of turmeric.
5. Mention favorable condition for development leaf blotch of turmeric .

Fig. 1: Leaf spot

Fig. 2: Leaf blotch

Plate-36: Photograph showing symptoms of major diseases of turmeric

37

Coriander Crop Diseases & Management

Name of Diseases	Causal Organism
Stem gall	*Protomyces macrospores* Unger
Powdery mildew	*Eysiphe polygoni* DC
Wilt	*Fusarium oxysporumf.sp. corianderii*

1). Stem Gall

Diagnostic Symptoms

- Symptom appears in the form of tumor-like swellings of leaf veins, leaf stalks, peduncles, stems as well as fruits.
- Infected veins show a swollen hanging appearance to the leaves.
- Initially the tumors are glossy which rupture later on and become rough. They are about 3 mm broad and up to 12.5 mm long.
- Badly affected plants may be killed.
- In the presence of excessive soil moisture, especially under shaded conditions, when the stem fails to harden and remain succulent, the tumors are numerous (Fig-1).

Etiology

Scientific Classification

Kingdom : Fungi

Division : Ascomycota

Class : Taphrinomycetes

Order : Taphrinales

Family : Protomycetaceae

Genus : *Protomyces*

Species : *macrospores*

- Mycelium: endophytic, intercellular, septate, and irregularly branched. After maturity, some of the hyphal cells swell, become ellipsoidal or globose, and thick walled.
- Cells so transformed represent the resting structures referred to as chlamydospores.
- *Chlamydospores:* Intercellular and multinucleate mycelium's are formed which are globose to ellipsoid, thick-walled, three-layered membrane and smooth with brownish colour. Mature Chlamydospores were multinucleate and measured 60-70 μm x 50-60 μm in diameter.
- Multinucleate protoplast of the chlamydospore migrates to the vesicle. A large vacuole develops in the centre of the vesicle displacing the multinucleate protoplasm to the peripheral region where it undergoes cleavage resulting in its many pieces.
- Protoplasm-pieces are "naked" and each one encloses within it a daughter nucleus.
- Naked cells so produced undergo reduction division that finally results in the formation of four spores inside each naked cell .
- On maturity, these ascospores collect in the centre of the vesicle which, however, bursts and releases them.
- Upon liberation, the ascospores may reproduce by budding in yeast-like manner or they may conjugate in pairs with one member of each pair continuing to bud.
- Conjugation cell (zygote) apparently germinates and forms a diploid mycelium that penetrates and enters into the host.

Disease Cycle

- Disease is soil borne.
- Pathogen may survive in soil as resting spore (chlaymydospores) for several years.
- Inoculum present in the soil are the source of primary infection.
- Secondary infection is carried by means of spores.

Epidemiology

- Relative humidity- >85%.
- Optimum Temperature – 15- 20^0 C.
- Ph- 7.5.

Management

- Deep summer ploughing of fields during summer.
- Use resistant or tolerant varieties.
- Soil solarization: Cover the beds with polythene sheet of 0.45 mm thickness for three weeks before sowing.
- Seed treatment with Carbendazim+ Mancozeb @ 2 gm /kg seed.
- Incubate Trichoderma @500 g in 100 Kg FYM for 15 days prior to its application. Inoculate Trichoderma @ 500 g in 100 Kg FYM /acre of the field.
- Field sanitation.
- Distroy the alternate host plants.
- Balanced applications of maures and fertilizers.
- Follow crop rotation with non host crops like cereals for 3 year.

Model Question Paper

Objective Type Questions

a. Choose the correct answer from the following

(1) Stem gall of coriander is caused by :

a. *Protomyces macrospores*

b. *Colletotrichum lindemuthianum*

c. *Pythium aphanidermatum*

d. *Sphacelotheca cruenta*

(2) Symptom appears in the form of tumor-like swellings of leaf veins, leaf stalks, peduncles, stems as well as fruits of coriander

a. Stem gall | b. Leaf spots

c. Powdery mildew | d. Fusarium wilt

(3) *Protomyces macrospores* belongs to order

a. Taphrinales b. Peronosporales

c. Helotiales d. Moniliales

(4) *Protomyces macrospores* belongs to class

a. Coelomycetes b. Taphrinomycetes

c. Pythiaceae d. Moniliaceae

(5) *Protomyces macrospores* cells transformed represent the resting structures referred to as

a. Chlamydospores b. Oospore

c. Ascospore d. Basidiospore

a Answer

S. No	Answer	S. No	Answer
1	a. *Protomyces macrospores*	4	b. Taphrinomycetes
2	a. Stem gall	5	a.Chlamydospores
3	a. Taphrinales	6	

(b) State whether the following statements are *True* or *False*

1. *Protomyces macrospores* is endophytic fungi .
2. Stem gall diseaseof coriander is soil borne in nature.
3. *Protomyces macrospores* may survive in soil as chlaymydospores for several years.
4. Inoculum of stem gall diseasepresent in the soil are the source of primary infection on coriander.
5. Secondary infection stem gall disease is carried by means of spores.
6. Relative humidity greater than 85% is congenial for development of stem gall of coriander.
7. *Protomyces macrosporus* belongs to family protomycetaceae.
8. *Protomyces macrosporus* has a complex life cycle including ascospores and chlaymydospore.
9. The pathogen *Protomyces macrosporus* survive in soil as resting spore called chlaymydospore.

10. Foliar spray with sulfur fungicide control stem gall disease of coriander.
11. The symptom appears in stem gall of coriander in the form of blister like swellings of leaf veins, leaf stalks, peduncles, stems as well as fruits.

b. Answer

S. No	Answer	S. No	Answer
1	True	6	True
2	True	7	True
3	True	8	True
4	True	9	True
5	True	10	True

Descriptive Questions

a. Long answer questions

1. Give a comprehensive understanding of the causal organism , symptoms, disease cycle, and management of stem gall of coriander.
2. Mention the causes, symptoms that can be identified most easily, disease cycle and integrated management of the stem gall of coriander.

b. Short answer Questions

1. List the integrated management strategies for Stem gall of coriander..
2. Explain epidemiological condition of Stem gall of coriander
3. Make some recommendations of new generation fungicide to control Stem gall of coriander
4. Which fungal pathogen causes Stem gall of coriander? Give its systematic position.
5. Write down the integrated management practices of Stem gall of coriander.

Fig. 1: Stem gall

Plate-37: Photograph showing symptoms of major diseases of coriander

38

Marigold Crop Diseases & Management

S. No	Name of Diseases	Causal Organism
1	Botrytis gray mould	*Botrytis cinerea*
2	Fusarium wilt	*Fusarium oxysporum*
3	Bacterial leaf spot	*Pseudomonas tagetis*
4	Leaf Spot	*Alternaria* spp

1) Botrytis Gray Mould

Diagnostic Symptoms

- Appearance of a gray, fuzzy mold leaves, stems and on flowers of or plants. Also called, "gray mold".
- Spotting or blight on leaf and flower stem lesions, and dieback.
- Flower parts become necrotic and die(Fig-1).

Etiology

Domain : Eukaryota

Kingdom : Fungi

Division : Ascomycota

Class : Leotiomycetes

Order : Helotiales

Family : Sclerotiniaceae

Genus : *Botrytis*

Species : *cinerea*

Telemorph : *Botryotinia fuckeliana*

- Colony: fast-growing, white, low, flaky, becoming grey to brownish grey.
- Conidia: ellipsoidal or ovoid, 6.1 to 8.5 × 5.1 to 9.8 μm and from PDA cultures were 5.1 to 10.5 × 4.3 to 7.2 μm .

- Conidiophores: straight or flexuous, septate, with an inflated basal cell, brown to light brown, and measured 105 to 425 × 9 to 28 μm.
- Sclerotia: Black ranging from 0.7 to 4.8 × 1.0 to 3.7 mm.

Disease Cycle

- Pathogen overwinter as sclerotia on dead plant debris. and as sclerotia or conidia in infested soil.
- Primary infection: Pathogen mycelial strands from previously infected plant parts can grow onto healthy plant parts and infect them.
- Secondary infection: Spores spread by wind or splashing water to infect dying, wounded, or extremely soft plant tissues in the spring.

Epidemiology

- Moist, humid environment is ideal for pathogen sporulation and spread.
- Optimum temperature for germination of conidia/spore & establishment of infection- 18 to 23^{0}C.
- Relative humidity for spore germination-90 to 100 percent.

Management

- Avoid splashing water on the foliage when watering.
- Remove and destroy all infected plant parts as soon as they are observed.
- Give adequate space between plants to allow for good air circulation.
- Avoid fertilizing with excessive amounts of nitrogen.
- Avoid unnecessarily wounding plants.
- Only blemish-free, nonsenescent flowers or plant material should be stored.
- Storage area should be clean, cool, and dry without free moisture on the walls, ceiling, or floor and with a humidity of 90 to 95 percent to prevent shrinking or shriveling of plant material. The temperature should be as close to freezing as possible.
- Foliar spray with Carbendazim + Mancozeb @ 0.2% or Hexaconzole + Zineb @ 0.2%,repeat after 15 days.
- Biological controls have been successfully used in the form of fungi-like *Trichoderma harzianium* Rifai to control gray mold.

Model Question Paper

Objective Type Questions

a. Choose the correct answer from the following

(1) Botrytis gray mould of marigold is caused by

a. *Botrytis cinerea*

b. *Colletotrichum lindemuthianum*

c. *Pythium aphanidermatum*

d. *Sphacelotheca cruenta*

(2) The telemorphic stage of botrytis gray mould of marigold is

a. *Botryotinia fuckeliana*

b. *Colletotrichum lindemuthianum*

c. *Sphacelotheca cruenta*

d. *Pythium aphanidermatum*

(3) *Botryis cinerea* is a type of fungus

a. Teleomorphic　　b. Anamorphic

c. Pseudomorphic　　d. Amorphic

(4) *Botrytis cinerea* belongs to order

a. Helotiales　　b. Peronosporales

c. Melanconiales　　d. Moniliales

(5) *Botrytis cinerea* belongs to class

a. Coelomycetes　　b. Taphrinomycetes

c. Pythiaceae　　d. Leotiomycetes

(6). *Botrytis cinerea* overwinter as on dead coriander plant debris

a. Chlamydospores　　b. Oospore

c. Sclerotia　　d. Basidiospore

Answer

S. No	Answer	S. No	Answer
1	a. *Botrytis cinerea*	4	a.Helotiales
2	a. *Botryotinia fuckeliana*	5	d. Leotiomycetes
3	b. Anamorphic	6	c.Sclerotia

(b) State whether the following statements are True or False

1. The perfect stage of *Botrytis cinerea* is *Botryotinia fuckeliana.*
2. *Botrytis cinerea* belongs to class Leotiomycetes.
3. Appearance of a gray, fuzzy mold on flowers, leaves or stems of marigold is called gray mold.
4. The colony of *Botrytis cinerea* is fast-growing and white.
5. *Gliocladium roseum* is a fungal parasite of *Botrytis cinerea.*
6. Different *Botrytis cinerea* strains show considerable genetic variability.
7. The conidia of *Botrytis cinerea* dispersed by wind and by rain-water, cause new infections.
8. *Botrytis cinerea* is a necrotrophic fungus that affects many plant species.
9. Excessive application of nitrogen will increase the incidence of disease while not improving yields.
10. *Botrytis cinerea* produces highly resistant sclerotia as survival structures in older cultures.

b. Answer

S. No	Answer	S. No	Answer
1	True	6	True
2	True	7	True
3	True	8	True
4	True	9	True
5	True	10	True

Descriptive questions

a. Long answer questions

1. Explain in detail the symptoms, etiology, disease cycle, epidemiology and management of botrytis gray mould of marigold.

2. Describe in detail the most distinguishing symptoms, causal organism, disease cycle and integrated management of botrytis gray mould of marigold.

b. Short answer questions.

1. Write down the etiology of pathogen causing botrytis gray mould of marigold.
2. Which fungal pathogen causes botrytis gray mould of marigold? Give its systematic position.
3. Write down the integrated management practices of botrytis gray mould of marigold.
4. Give the diagrammatic representation of the disease cycle of the botrytis gray mould of marigold.
5. Explain epidemiological condition of botrytis gray mould of marigold.
6. Make some recommendations for new generation fungicides to control botrytis gray mould of marigold.

Fig. 1: Botrytis grey mould

Plate-38: Photograph showing symptoms of major diseases of marigold

39

Rose Crop Diseases & Management

Name of Diseases	Causal Organism
Powdery mildew	*Sphaerothecapannosa var. rosae*
Die-back	*Diplodia rosarum*
Black spot	*Diplocarpan rosae*
Rust	***Phragmidum mucronatum***

1) Powdery mildew

Diagnostic Symptoms

- A white, powdery fungal growth on the upper, lower or both leaves and shoots.
- There may be discolouration of the affected parts of the leaf, and heavily infected young leaves can be curled and distorted.
- Mildew growth may also be found on the stems, flower stalks, calyces and petals.
- Heavily infected flower buds frequently fail to open properly.
- Mildew growth on stems and flower stalks is usually thicker and more mat-like than that on the leaves.
- Mildew growth on all parts may turn browner as it ages (Fig-1).

Etiology

Scientific Classification

Kingdom : Fungi

Division : Eumycota

Class : Leotiomycetes

Order : Erysiphales

Family : Erysiphaceae

Genus : *Sphaerotheca*

Species : pannosa

- Pathogen: *Sphaerotheca pannosa var. rosae/Sphaerotheca pannosa.*
- Obligate parasite: Pathogen must have a living host complete its life cycle, which can be as short as 72to 96 hours in favouable conditions.
- Mycelium: white, septate, ectophytic and sends globose haustoria into the epidermal cells of the host.
- Conidiophores: are short and erect.
- Conidia: one celled, oblong, minutely verrucose with many large fat globules.
- Cleistothecia: formed towards the end of the season on the leaves, petals, stems and thornswith simple myceloid appendages.
- Ascus: contains eight ascospores.

Disease Cycle

- Pathogen survives as mycelium in dormant buds and shoots or as cleistothecia.
- They discharge their spores in the spring.
- Secondary infection occurs through wind borne conidia throughout year.
- However, when it grows it can survive in buds, emerging with the appearance of the flowers.

Epidemiology

- Hot and warm days with cool nights.
- Water on leaf surface prevents the spores from germinating.

Management

- Collection and burning of fallen leaves.
- Prune the infected plants and dispose of them properly.
- Provide plants with adequate nutrients and water to maintain their immune difenses.
- Keep the soil well watered and mulched to prevent moisture loss and to cover up overwiterinng spores.

- Maintain proper spacing to provide good air circulation and prune them regularly to prevent overcrowding.
- Use fan to provide adequate ventilation during humid nights.
- Resistant varieties: Ashwini, Ambika, Angeles, American pride, Surabhi.
- Foliar spray with combination fungicide viz., Captan + Hexaconazole WP @ 0.3% or Azoxystrobin + Tebuconazole @ 0.1% or Tebuconazole with trifloxystrobin @ 0.1%, if required repeat after 15 days.

2) Die-back

Diagnostic Symptoms

- Browning and dieback of the tips of young shoots in spring.
- Browning and dieback of a pruning stub, which then progresses further down the branch.
- Dieback of twigs, branches, main stems or even the whole plant at any time of year.
- Fungal structures, such as tiny black fruiting bodies, are sometimes visible on the affected parts of the plant.
- In some instances there may also be associated root decay (Fig-2).

Etiology

Scientific Classification

Kingdom : Fungi

Division : Ascomycetes

Class : Dothideomycetes

Order : Botriosphaeriales

Family : Botriosphaeiaceae

Genus : *Diplodia*

Species : rosarum

- Pathogen: *Diplodia rosarum/Botryodiplodia theobromae.*
- Pathogen produces round, black pycnidia which bear spores.
- Pycnidia: round, black pycnidia which bear spores.

- Pycnidiospores: dark coloured and two celled.
- Perithecia: immersed in the host tissue and are surrounded by a pseudostroma.
- Ascospores: ellipsoidal or fusoid, hyaline, two celled with the septum in or near the middle.

Disease Cycle

- Pathogen survives inplant debris.
- Spores from pynidia are considered as primary source of infection and spread.
- Scondary infection takes place thorugh spores spread through air current or blowing rain.

Epidemiology

- High humidiy -40 -70%.
- Temperature-30-32^{0}C.

Management

- Plant roses in well-prepared soil, making sure that the roots are well spread out.
- Do not plant roses into soil that has grown roses previously, without taking remedial action. Change the soil to a depth of at least 30cm and a width of at least 60cm.
- Use resistant / tolerant varieties: Blue moon, Red gold, Summer queen, etc.
- Pruning should be done so that lesions on the young shoots will be eliminated and apply chaubatia paste in the pruned area.
- Avoid soils that are prone to either drought or waterlogging.
- Feed plants in spring with a proprietary rose fertiliser and mulch the soil to prevent water loss.
- Foliar spray with Captan + Hexaconazole WP @ 0.3% or Azoxystrobin + Tebuconazole @ 0.1% or Tebuconazole + Trifloxystrobin @ 0.1%, if required repeat after 15 days.

C) Black spot

Diagnostic Symptoms

- Dark brown tar coloured spots with fringed borders.
- Spots coalesce forming large patches.
- Infected leaves turn brown and defoliate.
- Fungus may also attack stems and flowers of rose bushes (Fig-3).
- On stems, infected areas are blackened with blistered appearance dotted withpustules.

Etiology

Scientific Classification

Kingdom : Fungi

Division : Ascomycota

Class : Leotiomycetes

Order : Helotiales

Family : Dermateaceae

Genus : *Diplocarpan*

Species : *rosae*

- Pathogen: *Diplocarpan rosae (Anamorph: Marssonina rosae).*
- Vegetative body of the fungus consists of two parts *viz*., the subcuticular mycelium and the internal mycelium.
- Fungus produces acervuli between the outer wall and cuticle of the epidermis.
- Fungus produces *Marssonina*-type conidia in acervuli and ascospores in tiny apothecia formed in old lesions.
- Conidiophore: short arise from a thin black stroma and give rise to successive crops of conidia.
- Conidia push up and rupture the cuticle.

Disease Cycle

- Pathogen survives as mycelia, ascospores and conidia in infected leaves and canes.
- Both kinds of spores can cause primary infections of leaves in the spring by direct penetration.
- The mycelium grows in the mesophyll, but within two weeks forms acervuli and conidia at the upper surface.
- Conidia are produced throughout the growing season and cause repeated infections during warm, wet weather.

Epidemiology

- Wet spring days and moist nights.

Management

- Use resistant / tolerant varieties: Bebebune, Coronado ,Grand opera, Sphinx.
- Affected parts should be collected and destroyed.
- Space roses far enough apart for good air circulation.
- Spray Tridemorph @0.025% or Benomyl @0.1% at weekly intervals starting with the sprouting of the plants till new foliage appears.
- Avoid overhead watering and keep foliage as dry as possible.
- Radiance - escape infection due to waxy surface.

Model Question Paper

a. Objective Type Questions

1. Powdery mildew of rose is caused by.......................... ..
    a. *Botrytis cinerea*
    b. *Colletotrichum lindemuthianum*
    c. *Spaerotheca pannosa var. rosae*
    d. *Sphacelotheca cruenta*

2. Black spot of rose is caused by ....

   a. *Diplocarpan rosae*

   b. *Colletotrichum lindemuthianum*

   c. *Pythium aphanidermatum*

   d. *Sphacelotheca cruenta*

3. The mycelium *of Diplocarpan rosae* forms structure at the upper surface within two weeks of infection

   a. Chlamydospores b. Pycnidia

   c. Acervuli d. Basidiospore

4. Foliar spray with sulfur fungicide control which disease of rose

   a. Stem gall b. Leaf spots

   c. Powdery mildew d. Fusarium wilt

5. The *Diplocarpan rosae* produces *Marssonina*-type conidia in acervuli and ascospores in tiny fruiting body called

   a. Apothecia b. Pycnidia

   c. Acervuli d. Perithecia

6. The telemorphic stage of *Diplocarpan rosae is*

   a. *Marssonina rosae*

   b. *Colletotrichum lindemuthianum*

   c. *Pythium aphanidermatum*

   d. *Sphacelotheca cruenta*

7 The optimum temperature for development of dieback of rose is

   a. 8-12 ^{0}C b 15-20 ^{0}C.

   c. 16-26 ^{0}C d. 30-32 ^{0}C

8. A fruiting body called cleistothecia formed towards the end of the season on the leaves, petals, stems and thorns of rose with simple myceloid appendages due to infection of

   a. *Sphaerothecapannosa var. rosae* b. *Colletotrichum lindemuthianum*

   c. *Pythium aphanidermatum* d. *Sphacelotheca cruenta*

9. *Sphaerotheca pannosa var. rosae is*
   a. Obligate parasite
   b. Facultative parasite
   c. Obligate saprophyte
   d. Facultative saprophyte

10. *Sphaerothecapannosa var. rosae* survives in soil as resting strucuture called
   a. Apothecia
   b. Pycnidia
   c. Cleistothecia
   d. Perithecia

(a) Answer

S. No	Answer	S. No	Answer
1	a.*Sphaerotheca pannosa var. rosae*	6	a.*Marssonina rosae*
2	a.*Diplocarpan rosae*	7	d.30-32^0 C
3	c.Acervuli	8	a.*Sphaerothecapannosa var. rosae*
4	c.Powdery mildew	9	a. Obligate parasite
5	a.Apothecia	10	c.Cleistothecia

b. State whether the following statements are True or False

1. *Diplodia rosarum* produces round, black pycnidia which bear spores.
2. *Diplocarpan rosae* survives as mycelia, ascospores and conidia in infected leaves and canes.
3. *Diplocarpan rosae* produces *Marssonina*-type conidia in acervuli and ascospores in tiny apothecia formed in old lesions of rose.
4. The pathogen *Sphaerotheca pannosa* is an obligate parasite.
5. White floury patches symptom appear on both side of the leaves of rose in powdery mildew disease.
6. The fungus *Diplocarpan rosae*produces acervuli between the outer wall and cuticle of the epidermis.
7. Spores from pycnidia are considered as primary source of infection and spread of *Diplocarpan rosae*.
8. Scondary infection of die-backtakes place thorugh spores spread through air current or blowing rain.
9. *Sphaerotheca pannosa* ascus contains eight ascospores.
10. Foliar spray with combination fungicide viz., Captan + Hexaconazole WP @ 0.3% control powdery mildew of rose.

b Answer

S. No	Answer	S. No	Answer
1	True	6	True
2	True	7	True
3	True	8	True
4	True	9	True
5	True	10	True

Long answer questions

1. Explain in detail about symptoms, etiology, disease cycle, epidemiology and management of powdery mildew of rose .
2. Briefly describe the diagnostic symptoms, etiology, disease cycle, epidemiology and management of dieback diseases of rose.
3. Describe the black spot of rose in the following headings:

   (i) Pathogen (ii) Symptoms

   (iii)Disease cycle (iv) Disease management
4. Describe in detail the most distinguishing symptoms, causal organism, disease cycle and integrated management of die back of rose.
5. Give the diagrammatic representation of the disease cycle of the following diseases of rose

   (i) Powdery mildew (ii) Die back

   (iii)Black spot

Section 'B'

Short answer questions

1. Write down the etiology of pathogen causing powdery mildew of rose.
2. Which fungal pathogen causes powdery mildew of rose? Give its systematic position.
3. Write down the integrated management practices of black spot diseases of rose.
4. Suggest some chemicals for management of powdery mildew of rose.
5. Describe favourable condition for development of black spot and dieback disease of rose.

Fig. 1: Powdery mildew

Fig. 2: Die back

Fig. 3: Black spot

Plate-39: Photograph showing symptoms of major diseases of rose